Graphene and Graphene-Based Materials

Structure and Mechanical Properties

Graphene and Graphene-Based Materials
Structure and Mechanical Properties

Julia Baimova

Institute for Metals Superplasticity Problems,
Russian Academy of Sciences, Russia

World Scientific

NEW JERSEY · LONDON · SINGAPORE · BEIJING · SHANGHAI · HONG KONG · TAIPEI · CHENNAI · TOKYO

Published by

World Scientific Publishing Co. Pte. Ltd.

5 Toh Tuck Link, Singapore 596224

USA office: 27 Warren Street, Suite 401-402, Hackensack, NJ 07601

UK office: 57 Shelton Street, Covent Garden, London WC2H 9HE

Library of Congress Control Number: 2025026917

British Library Cataloguing-in-Publication Data
A catalogue record for this book is available from the British Library.

GRAPHENE AND GRAPHENE-BASED MATERIALS
Structure and Mechanical Properties

ISBN 978-981-98-1242-4 (hardcover)
ISBN 978-981-98-1243-1 (ebook for institutions)
ISBN 978-981-98-1244-8 (ebook for individuals)

For any available supplementary material, please visit
https://www.worldscientific.com/worldscibooks/10.1142/14291#t=suppl

Desk Editor: Shaun Tan Yi Jie

Typeset by Stallion Press
Email: enquiries@stallionpress.com

Preface

The subject matter of this publication is the analysis of the structure and mechanical properties of graphene and graphene-based nanomaterials. Numerous aspects of the mechanical properties are considered, including deformation behavior, tensile strength, hardness, and deformation mechanisms for a wide variety of carbon structures. Researchers unfamiliar with the variety of carbon can find an overview of its different structures and properties. For more experienced practitioners, these chapters serve as a concise handbook describing the most important properties.

Chapter 2 (Introduction) is devoted to the description of the structural features of carbon polymorphs, graphene and different graphene-based materials. Synthesis methods, new carbon structures and their properties are briefly described. Three main classes of the carbon structures are considered: (i) carbon polymorphs including graphene, (ii) graphene-based materials and separately (iii) graphene-based composites.

Chapter 3 provides a comprehensive description of the mechanical properties of different carbon structures, including graphene, graphynes, graphene-based nanostructures, and carbon/metal composites. Different aspects of mechanical properties are considered – tensile strength, elastic constants, deformation mechanisms.

Chapter 4 covers the molecular dynamics methodology for the analysis of different mechanical properties. Understanding of the application of this method allows for a better understanding of the mechanical properties and the relationship between properties and structure.

Chapter 5 deals with the application of carbon nanostructures in different fields such as electronics, protective coatings, and composite technologies, to name a few. Future prospects of the application of graphene and graphene-based materials are described with the examples.

The completion of this book would not have been possible without the excellent teamwork of the members of our research group, with whom the discussions were always fruitful. I greatly appreciate the efforts of Radik Mulyukov, Sergey Dmitriev, Karina Krylova, Angelina Akhunova, Ramil Murzaev, Liliya Safina, Polina Polyakova, Elizaveta Rozhnova, and Elena Karpinskaya, who provided valuable feedback during the preparation of this book. We are also deeply grateful to our collaborators who have provided us with tremendous support in conducting research on molecular dynamics simulation. Last but not least, I would like to thank my mother, whose love and devotion have prepared me for a lifetime of success.

I also acknowledge the financial support from the State Assignment of the Institute of Metals Superplasticity Problems, Russian Academy of Sciences.

Julia A. Baimova

About the Author

Dr Julia Baimova, PhD DSc is the youngest Professor of the Russian Academy of Sciences (RAS), where she is Head of Laboratory of Physics and Mechanics of Carbon Nanomaterials at the Institute for Metals Superplasticity Problems (IMSP), RAS. She is a specialist in the field of physical and mechanical properties of carbon materials, and has published over 100 peer-reviewed papers in international journals. She is Deputy Editor-in-Chief of *Letters on Materials*, Associate Editor of the *Journal of Micromechanics and Molecular Physics*, Editor of *Materials* (MDPI), and an elected Member of the Russian National Committee of Theoretical and Applied Mechanics.

Contents

Chapter 1

Introduction

1.1 Carbon Atom

Carbon (C) is one of the most unique elements in the periodic table. It has been known since the beginning of human history [259]. In Mendeleev's periodic table it has an atomic number 6, and atomic mass 12.011. The nuclei of the carbon isotope ^{12}C consist of 6 protons and 6 neutrons. In 1961, the International Union of Pure and Applied Chemistry chose the carbon atom as the mass unit: 1 atomic mass unit is defined as 1/12 the mass of ^{12}C. The isotope ^{13}C is radioactive.

The neutral carbon atom ^{12}C with 6 protons and 6 neutrons has two electrons in the $1s$ atomic orbital, and four other electrons are in the $2p$ orbital. Due to this electronic structure, carbon atoms exhibit a variety of valence states and types of hybridization during the formation of covalent bonds (sp, sp^2, sp^3 hybridization), which further results in the formation of numerous carbon allotropes. Carbon differs from all the other elements in its ability to bond with other atoms to form materials with an extraordinary range of chemical and physical properties. Because of its allotropy, carbon can exist in diverse solid forms such as soot, graphite, diamond, fullerenes, nanotubes, and one-atom-thick layers of carbon. In combination with the other elements, it can form a nearly infinite number of compounds.

Carbon–carbon (C–C) bonds are formed by one σ bond, and double (C=C) or triple (C≡C) bonds by the combination of one σ bond and either one or two π bonds. A very detailed explanation of the electronic structure can be found in [217].

Figure 1.1 presents the four hybrid sp^3 orbitals for the methane molecule CH_4. Each lobe of the four orbitals is aligned along the four corners of a tetrahedron. In this case, we have four equivalent C–H covalent bonds and

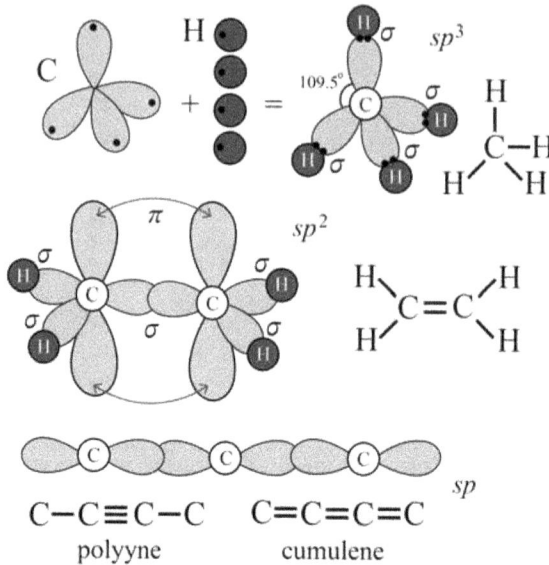

Fig. 1.1. Carbon structures with different hybridization: methane molecule CH_4, sp^3 hybridization; ethene molecule C_2H_4, sp^2 hybridization; two variants of the carbyne chain, sp hybridization.

four equivalent (109.5°) H-C-H covalent angles. In ethene C_2H_4, there is a single σ bond between two C atoms, and each carbon atom forms another two σ bonds with H atoms. The unhybridized atomic orbitals overlap with each other side by side, and they form a π bond. The carbyne chain has sp hybridization for both configurations: polyyne and cumulene, although these structures differ by the double (C=C) or triple (C≡C) bonds.

The C-C interatomic bond in graphene consists of bonds formed by electron clouds aligned along the axis of the bond (σ bonds) and bonds formed by electron clouds aligned perpendicular to both the bond axis and the graphene plane (π bonds). The σ bonds require a lot of energy to break; these bonds are responsible for the extraordinary strength of graphene. The electrons in the π bonds, on the other hand, are very weakly bound and are free to move at great velocity through the graphene layer. These electrons are responsible for the extremely high electrical conductivity of graphene.

Carbon materials, and especially graphene and graphene-based materials, demonstrate unusual physical, mechanical and electronic properties. The carbon atom can form structures of various dimensions, namely 3D (diamond, lonsdaleite), 2D (graphene, graphyne), 1D (carbyne, carbon nanotubes) and even 0D (fullerene, graphene quantum dots). To date, numerous different carbon polymorphs have been predicted and synthesized, even carbon-carbon composites. All differ in their atomic structure, size, dimensionality, morphology and exhibit a wide variety of different properties. Each of these structures can be characterized by certain unique properties: the ultra-high hardness of diamond, the unique electrical and thermal conductivity of graphene, the outstanding strength of graphene and carbon nanotubes (CNTs), and the abnormally high melting temperatures, to name a few.

1.2 Carbon Polymorphs

1.2.1 *Nanostructures*

Among carbon polymorphs, there are three very famous nanostructures – fullerene, graphene and CNT. To date, various fabrication methods have been developed to successfully synthesize these polymorphs which open up new opportunities for their future applications.

As widely known, nanostructures exhibit unique properties that are very different from conventional structures such as graphite and diamond. Kittel [125] formulated the reasons the structure of nanomaterials significantly affects their properties:

(a) most of the atoms in the nanomaterial belong to the surface, in contrast to bulk materials, where the proportion of surface atoms is small;
(b) the ratio between the surface energy and the total energy can be very close;
(c) the electron mean free path is limited by geometric dimensions;
(d) the wavelength or boundary conditions can lead to the phenomena of optical adsorption.

The story of nanomaterials starts in 1959, when the idea of nanostructured materials was unknown to society. The famous Nobel lecture "Plenty of Room at the Bottom" given by Richard Feynman became a classic, innovative talk on nanotechnology [72]. Feynman stated that it is possible to manipulate matter on an atomic scale. In his lecture he discussed that it should be possible to create nanoscale machines where atoms are arranged as we want. Since then, the idea of nanotechnology has come a long way in the whole.

Many prospects for the development of the nanomaterials industry are currently concentrated around graphene and graphene-based nanostructures. Although such structures have gained the greatest popularity since the experimental exfoliation of graphene in 2004, other novel carbon structures have also attracted much attention from scientists in recent decades. At present, we can already talk about the whole field of carbon nanomaterials as an established branch of science. The structure and properties of various carbon polymorphs and their modifications, synthesis, and properties gained a lot of attention during the last 20 years. Figure 1.2 presents the main milestones in the synthesis of carbon polymorphs and other structural modifications.

1.2.2 *Brief History of Carbon Polymorphs*

A third basic polymorph of carbon after diamond and graphite was synthesized in 1960: a 1D carbyne chain [118, 255]. Soviet chemists A. M. Sladkov, V. V. Korshak, V. I. Kasatochkin and Yu. P. Kudryavtsev discovered a linear form of carbon, which they called carbyne (from the Latin carboneum (carbon) with the ending "yne", used in organic chemistry to denote the acetylene bond). Carbyne is an infinite carbon chain that can exists in two forms: polycumulene ($=C=C=C=C=$)$_n$ (containing only double bonds) and polyyne ($-C\equiv C-C\equiv C-$)$_n$ (containing alternating single and triple bonds). In 1968, the carbyne-like carbon was found in geological rocks formed in a meteorite crater [65]. After this discovery, the history of carbon nanopolymorphs began.

In 1985, a new form of carbon was discovered — fullerene, which is a molecule composed of 60 carbon atoms in the shape of an icosahedra or a soccer ball [135]. First-known fullerene molecules were C_{60} and C_{70}, while to date we can obtain fullerene cages with number of atoms from 20 to 960. For the discovery of fullerenes, Kroto, Smalley, and Curl were awarded the Nobel Prize in Chemistry in 1996. However, this discovery had a long history: the first work describing the C_{60} molecule was published in 1970 by the Japanese scientist E. Osawa, who suggested the stability of C_{60}; the following year, in his joint book with Z. Yoshida, a more detailed description of the various aromatic properties of this molecule appeared. A theoretical study of this problem was carried out by Russian scientists D. A. Bochvar and E. G. Galpern in 1972 [26]. Earlier in 1966, D. Jones suggested that the introduction of five-atom defects into a graphite layer consisting of regular six-atom rings could transform this flat layer into a hollow closed shell.

In 1991, the Japanese scientist Iijima succeeded in discovering CNTs [110]. However, in the first studies, multi-walled CNTs were discovered,

History of carbon materials

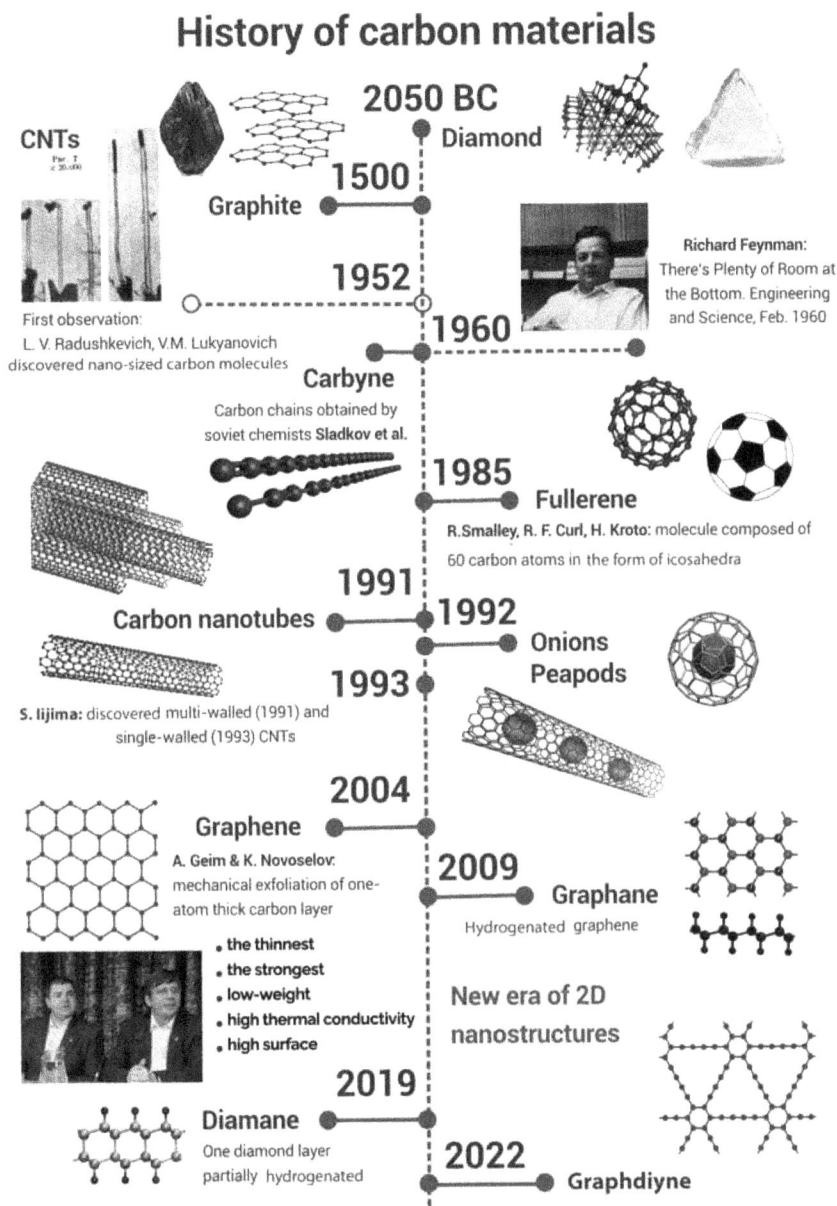

Fig. 1.2. History of carbon materials. Photo of R. Feynman reprinted with permission from the Internet under Creative Commons license. First observation of CNT reprinted with permission from [222]. Photo of A. Geim & K. Novoselov reprinted with permission from https://en.wikipedia.org/wiki/Discovery_of_graphene under Creative Commons license.

and only later, in 1993, single-walled CNTs were obtained, which led to the active development of research in this field. Researchers Dresselhaus and Iijima were also awarded a prestigious prize for their discovery — the Cavli Prize. It should be noted that the first time CNTs were observed was by Soviet chemists L. V. Radushkevich and V. M. Lukyanovich in 1952. They published an article in the *Russian Journal of Physical Chemistry* entitled "On the structure of carbon formed during the thermal decomposition of carbon monoxide on an iron contact" with photographs taken with a transmission electron microscope, showing clusters of long molecules. Their length reached 5–7 μm, and the diameter of the thinnest was about 30 nm. The results were published in Russian, which may explain why the related papers are not well known and cited [222].

Graphene was next and was first mechanically exfoliated in 2004 by Geim and Novoselov [189]. However, the word "graphene" itself and the definition of the material were introduced in 1962 by the German chemist Hans-Peter Boehm.

The main reason graphene was first exfoliated only 20 years ago is that the growth of 2D crystals is a very complicated process. The stability of 2D structures has been discussed for years (Fig. 1.3), especially for crystal growth at high temperatures, where thermal fluctuations affect the stability of the microstructure. The instability of 2D structures at finite temperatures was first discussed by Landau and Peierls [142, 202] and further by Mermin [169]. According to the so-called Mermin-Wagner theorem [169], long-wavelength fluctuations destroy the long-range order of 2D crystals. One way to achieve stability of the 2D material is a ripple formation [185], since single-layered graphene can spontaneously develop thermal ripples at finite temperatures [54, 70, 248]. Another way is to grow the 2D crystal on top of a 3D crystal, such as a metal surface or SiO_2, and further remove the bulk part. A third way is to split a layered material like graphite into individual atomic layers [82, 83]. Graphene exfoliation provides high structural and electronic quality, although this method is very delicate and graphene cannot be obtained in large amounts. In contrast, during graphene growth on top of the other crystal, graphene remains bound to the substrate, but it can considerably affect the graphene properties. Furthermore, the substrate can be removed to obtain a pure 2D layer, but it is quite complicated to obtain a one-atom-thick layer [124, 170, 194].

In the 1960s, new graphite intercalation compounds with higher basal plane conductivity were discovered [265, 270]. It was shown by *ab initio* calculations that graphene is thermodynamically unstable when its size

If graphene can exist

Rudolf Peierls

1934 •------•

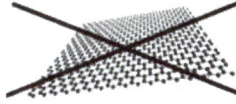

Landau-Peierls instability:
2D structures are unstable due to
the thermal fluctuations

Lev Landau

1937 •------•

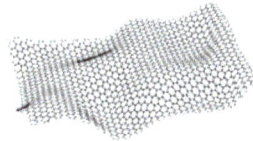

D. Mermin

1966 •------•

Mermin-Wagner theorem:
long-range magnetic order
does not exist in 2D

**The era of 2D
nanostructures**

Richard Feynman:
There's Plenty of
Room at the Bottom

•------• **1960**

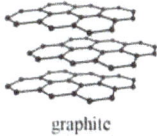 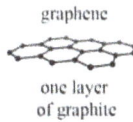

graphene

one layer
of graphite

graphite

•------• **1962**

Hanns-Peter Boehm: termed graphene;
first observation of graphene

2004

A. Geim & K. Novoselov:
mechanical exfoliation of
one-atom thick carbon layer

Fig. 1.3. How was it discovered that 2D structures can be stable? Photos reprinted with permission from the Internet under Creative Commons licenses.

is smaller than about 20 nm [247]. Various attempts have been made to synthesize graphene, including using the same approach for the growth of CNTs [131], such as chemical vapor deposition on metal surfaces [143, 183]. The idea of separating a single layer of graphite with scotch tape came from Oleg Shklyarevsky, who used tape to clean the surface of bulk graphite. Graphite particles remain on the adhesive tape, from which a single-layer graphene sample is then obtained by dissolving the tape. Geim and Novoselov were awarded the Nobel Prize in Physics in 2010 for their work on graphene.

1.2.3   *Classification of Carbon Nanostructures*

To date, there are several different classifications of carbon nanostructures; the most common are based on their electronic properties (hybridization) and dimensions (Fig. 1.4). Note that there is no agreement on how many allotropes of carbon exist, since new carbon nanostructures appear from time to time.

A classification of carbon structures based on atomic hybridization was proposed in [112, 227], and the study presented a classification by the number of neighboring atoms in the first coordination sphere or by the ratio of atoms forming 2, 3 or 4 covalent bonds in the material. Figure 1.5 presents the variety of carbon nanostructures with different hybridization.

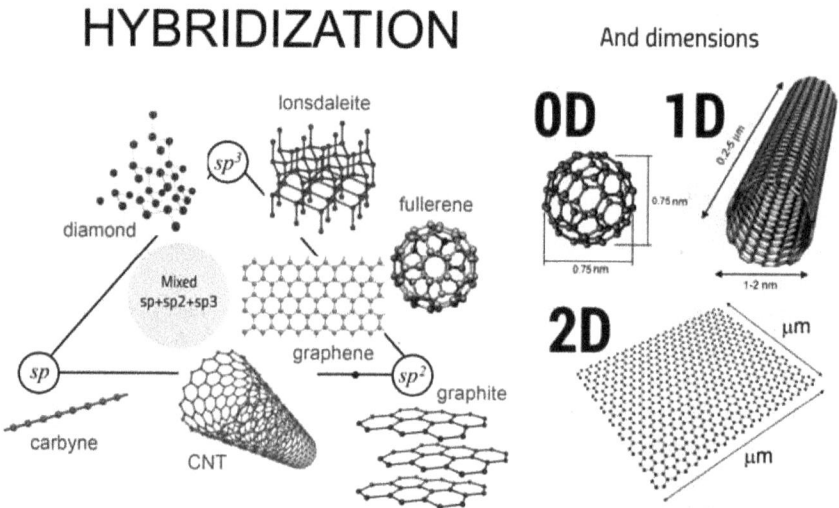

Fig. 1.4.   Different classifications of carbon nanostructures.

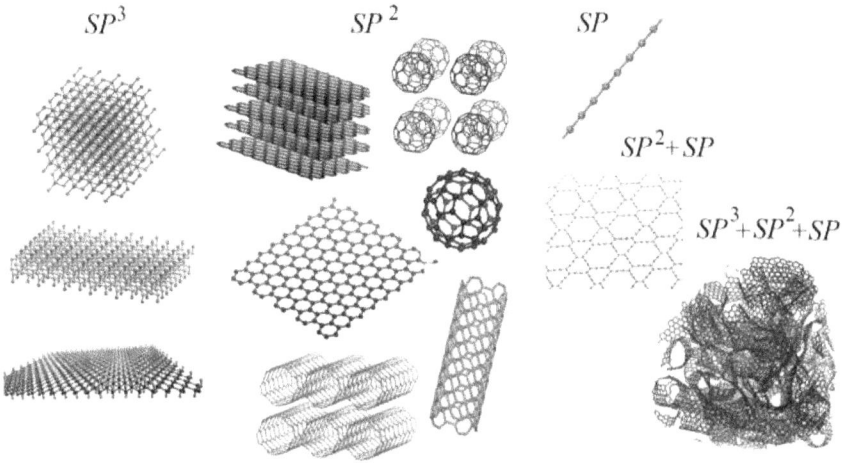

Fig. 1.5. Carbon structures with different hybridization: sp^3 diamond, diamane, graphane; sp^2 graphite, graphene, fullerene, fullerite, CNT and CNT bundle; sp carbyne; mixed hybridization — graphyne and graphene aerogel.

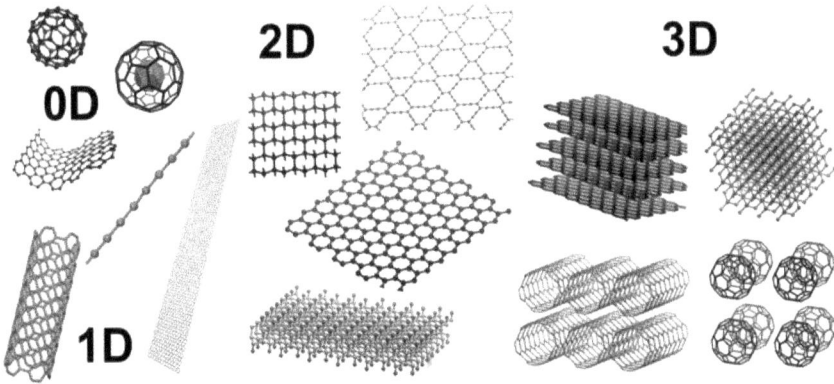

Fig. 1.6. Different carbon polymorphs according to their dimension. 0D: fullerene, onion, graphene dot. 1D: CNT, carbyne, graphene nanoribbon. 2D: graphene, diamane, graphyne. 3D: graphite, diamond, fullerite, CNT bundle.

Each of three main valence states can characterize the carbon polymorphs: sp^3 (diamond), sp^2 (graphite) and sp (carbyne). All other forms of carbon structures are transitional forms and can have both mixed hybridization and hybridization corresponding to one of the main structures.

Figure 1.6 shows molecular models of carbon nanostructures according to their dimensions. As can be seen, there is a large variety of structures in each of the dimensions, and the figure does not show all possible configurations: 0D (fullerene, onion, graphene dot), 1D (CNTs, carbyne, graphene nanoribbon), 2D (graphene, diamane, graphyne), 3D (graphite, diamond, fullerite, CNT bundle).

Chapter 2

Structure, Synthesis, and Properties Overview

2.1 Graphene and Other Low-Dimensional Carbon Polymorphs

To date there is a wide variety of graphene nanopolymorphs which can differ considerably by their structure, physical and mechanical properties. These new low-dimensional carbon polymorphs can be obtained directly from graphene by the combination of graphene flakes to more complex structures, by graphene functionalization, or by the changing of graphene morphology.

Figure 2.1 presents a small variety of low-dimensional carbon polymorphs: featured at the top are 2D polymorphs (diamane, graphane, graphdiyne) and the rest are quasi-2D structures - diamond-like phases (DLPs), graphene aerogels (crumpled graphene and honeycomb) and graphene-based composites. All of the presented structures have already been synthesized. Only these structures are presented because in Chapters 3 and 4, these nanostructures will be discussed from the point of view of their mechanical properties.

In the present chapter, the structure and some properties of fullerenes, carbon nanotubes (CNTs) and other similar structures will be discussed for better understanding of the variety of graphene derivatives. In one of the pioneering works on graphene, this structure was named "the mother of all graphites" [83]. The reason is that other carbon polymorphs can be somehow obtained from graphene: graphite is multi-layer graphene, CNT is graphene wrapped to a nanotube, and fullerene can be obtained from graphene by the introduction of 5-atom rings, to name a few.

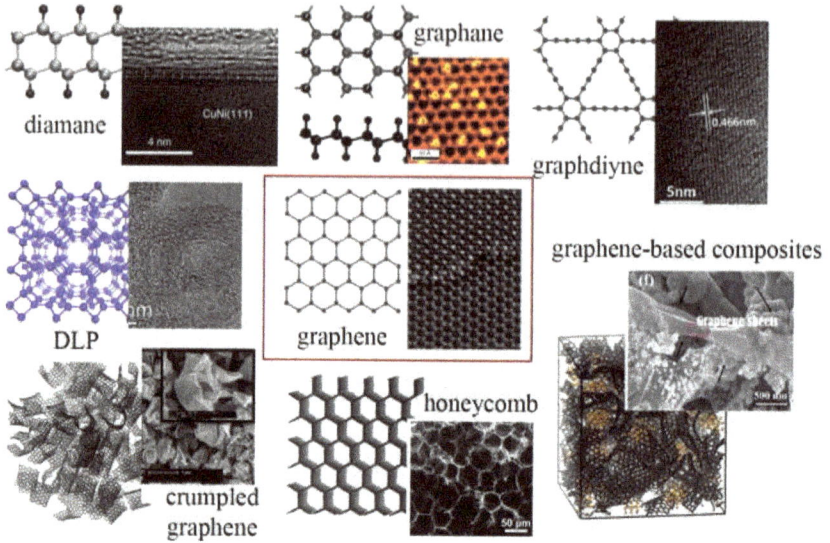

Fig. 2.1. Graphene and graphene derivatives.

2.1.1 *Graphene*

2.1.1.1 *Structure and Synthesis*

Graphene, one of the allotropes of elemental carbon, is a planar monolayer of carbon atoms arranged in a 2D honeycomb lattice with a carbon-carbon bond length of $a_0 = 0.142$ nm. Graphene can also be considered as one layer of graphite. Graphene is a very light material: its weight is only 0.77 mg/m^2, and its specific surface area is 2600 m^2/g.

A primitive unit cell for graphene consists of two atoms due to the hexagonal structure. The lattice vectors a_1 and a_2, joining equivalent points in adjacent unit cells, can be defined as:

$$a_1 = \left(\frac{\sqrt{3}a}{2}; \frac{a}{2}\right), \quad a_2 = \left(\frac{\sqrt{3}a}{2}; -\frac{a}{2}\right),$$

where $|a_1| = |a_2| = a = \sqrt{3}a_0$.

Graphene has been known since the 17th century, but it was experimentally exfoliated only in 2004. In the few years that have passed since the first publication on the fabrication and study of individual graphene flakes, so many different approaches to the synthesis of this material have been developed that it is surprising that graphene was not discovered many decades earlier. As an explanation for this mystery, it can be noted that when obtaining graphene, the most difficult stage is not so much the synthesis of samples, but the identification and determination of their main parameters (size, number of layers).

Two highly symmetric directions can be distinguished for graphene, the armchair and the zigzag (see Fig. 2.2b), along which the graphene properties can differ considerably.

There are currently many different methods for graphene fabrication. For example, growing graphene from SiC is one of the promising ways to produce large-area graphene samples for use in electronics. Another effective approach to the problem of separating graphite layers is the use of chemical oxidants. The most commonly used method for producing graphene is chemical vapor deposition (CVD), which has become very popular due to its simplicity and relatively low cost. This method is based on the possibility of

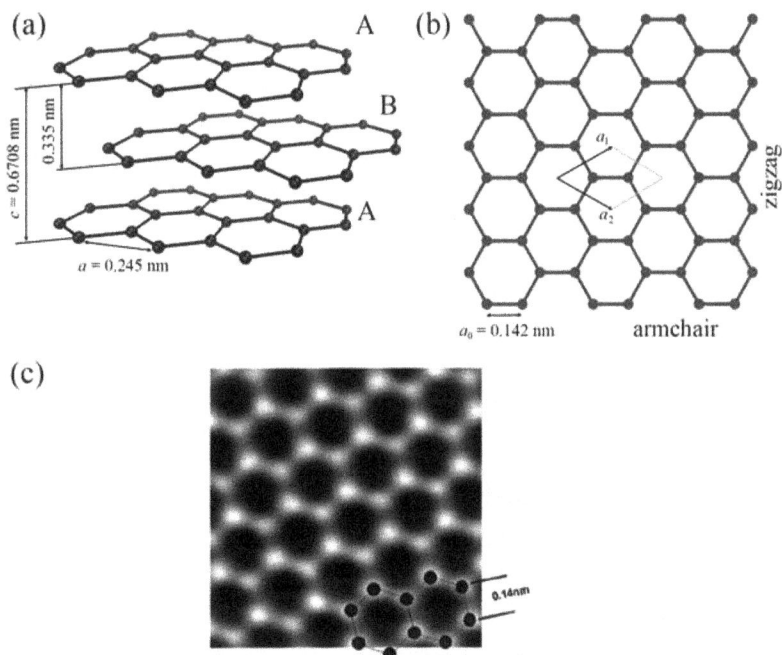

Fig. 2.2. (a) Atomic structure of Bernal (ABAB) stacked graphite. (b) Graphene lattice. (c) An atomic resolution image of a clean and structurally perfect graphene. Reprinted with permission from [272].

catalytic decomposition of gaseous hydrocarbons on the surface of certain metals to form various nanocarbon structures.

Each of the methods for the fabrication of the single-layer graphene has its own pros and cons. For example, strong interaction with the substrate can lead to a significant redistribution of charges between the materials; mechanical instability during transfer to plastic (for flexible electronics) can lead to destruction of the graphene lattice; the formation of external defects due to heterogeneous growth can reduce the electron mobility. At present, large graphene samples with sufficient electron mobility can be obtained using the CVD method, which means that the barrier to industrial graphene production can soon be overcome. In addition, the possibilities of growing carbon structures and controlled production of graphene with specified properties are being actively studied.

The production of graphene by micromechanical exfoliation is a simple process for splitting crystalline graphite. In this approach, graphene layers are separated from crystalline graphite either by rubbing small graphite crystals against each other or by using adhesive tape. In this case, the top layer of high-quality graphite is removed using a piece of adhesive tape, which, together with the graphite crystals, is pressed onto the selected substrate or is split with acid. If the adhesion of the lower graphene layer to the substrate exceeds the adhesion of the graphene layers to each other, the graphene layer can migrate to the surface of the substrate. As a result, using this amazingly simple procedure, graphene crystals of very high quality are obtained. This method works with almost any surface that demonstrates sufficiently good adhesion to graphene.

In the first experiments, mechanical exfoliation gave extremely low quantity of graphene, and in order to find a micron-sized graphene flake, it was necessary to examine large areas of the surface. Obviously, searching using traditional microscopy methods, such as atomic force or scanning electron microscopy (SEM), is an almost impossible task; in practice, its implementation is only possible using optical electron microscopy. Studies show that such an approach allows one to isolate single-layer graphene sheets with an ordered structure and having a width of about 10 μm and a length of about 100 μm. It is obvious that obtaining graphene by mechanical exfoliation presents significant difficulties for its mass production.

The main problem that arises when obtaining graphene by mechanical exfoliation is the difficulty of identifying them. As a result of microme-chanical action, a significant number of fragments are formed, which are graphene samples with a different number of layers (from one to a hundred).

The proportion of single-layer samples in this conglomerate is relatively small, so the main difficulty is associated with the detection of such single-layer samples.

The simplest method of exfoliating graphite into individual graphene layers is based on the use of surface-active organic liquids. The appearance of foreign atoms leads to an increase in the distance between the layers in graphite and, accordingly, to a decrease in the interaction energy between them. As a result, it becomes possible to exfoliate graphene layers under mechanical treatment. Long-term ultrasonic treatment of finely dispersed graphite and a surface-active liquid leads to the formation of a suspension containing single-layer graphene sheets, as well as graphene samples consisting of a small number of layers.

The most common method for graphene fabrication is CVD, which has become very popular due to its simplicity and relatively low cost. This method is based on the possibility of catalytic decomposition of gaseous hydrocarbons on the surface of some metals to form various nanocarbon structures. Currently, CVD has been used to obtain large amounts of graphene with sufficient electron mobility. An example of the successful use of the CVD method for graphene synthesis is a work in which a nickel film less than 300 nm thick was used as a substrate acting as a catalyst [290]. Graphene with different numbers of layers are clearly visible on the samples observed using scanning electron microscopy. Observations have shown that the average number of graphene layers and the degree of substrate coverage are determined by the thickness of the nickel film and the duration of the growth process. As is known, an arc discharge with graphite electrodes, occurring in an inert gas atmosphere, is one of the most effective ways of converting crystalline graphite into surface carbon structures. Based on this approach, methods for obtaining fullerenes and CNTs in macroscopic quantities were also developed.

In addition to SiC single crystals, metal surfaces with a well-ordered metallic structure, such as Ru(0001), are also successfully used as a substrate for epitaxial growth of graphene. The approach to layer-by-layer growth of large-area graphene on the Ru(0001) surface is based on the increasing temperature dependence of the solubility of carbon in transition metals. At high temperatures (1150°C), where the solubility of carbon is high, the metal sample is saturated with carbon by volume. Slow cooling of the sample to approximately 825°C leads to an approximately sixfold decrease in solubility, resulting in the release of excess carbon on the metal surface, and the surface is covered with extensive islands of carbon film.

Observations show that the film is single-layer epitaxial graphene sheets with a Moiré structure.

Not only monolayer graphene is often obtained, but also bilayer graphene, which is very different from the monolayer in its properties, and even has better properties for some applications. Two layers of graphene, stacked together at a distance of 0.34 nm, do not tend to be located exactly one above the other so that each carbon atom has a partner in the adjacent layer. Instead, bilayer graphene is mainly in the so-called Bernal A-B stacking state in the form of close packing: some of the atoms of the upper layer are projected directly into the centers of the hexagons of the lower layer (see Fig. 2.2a). Graphene layers are stacked in graphite in exactly the same way.

When infinite graphene crystals become finite, surfaces and boundaries appear, and when the size is on the order of several nanometers along one dimension and much larger along another, we have a graphene nanoribbon (GNR), which again differs from graphene in its properties. As with graphene, the main methods for obtaining GNR are CVD, chemical synthesis, obtaining nanoribbons from CNTs, scanning tunnel lithography, etc. The edges of the GNRs are chemically active, which allows for further modification of their electrical, chemical and magnetic properties by doping the edges with various atoms or molecules.

GNRs, like an infinite graphene, have three types of edge orientation: armchair, zigzag and chiral. Since GNR is a 1D structure (the width of the nanoribbon is much smaller than its length), the properties of the nanoribbons can differ greatly. Density functional theory (DFT) calculations showed that armchair GNR can have three band gap values depending on the width of the nanoribbon, and the gap width decreases with decreasing nanoribbon width. In addition, the number of layers also affects the electronic properties; for example, a two-layer armchair GNR has a band gap width smaller than a single-layer nanoribbon, and can behave both as a metal and as a semiconductor, while a single-layer armchair GNR exhibits purely semiconductor properties.

Zigzag GNRs are semiconductors with energy gap decreasing with increasing GNR width, with each individual edge atom playing an important role. For example, zigzag GNR with 16 edge atoms is a semiconductor with a gap of 0.8 eV, while the gap of zigzag GNR with 17 edge atoms decreases to 0.2 eV. Accordingly, the region of zero conductivity is reduced, and a very small external electric field can induce electron transfer through one channel. For zigzag GNR, quantum transport is determined by the edge states, which are expected to be spin-polarized. Indeed, the topologically zigzag edges give rise to special extended electron states.

The planar shape of graphene is not a stable configuration, as was previously shown by Landau and Peierls [142, 202] and further by Mermin [169] (see Chapter 1). It has been reported that graphene loses its planar shape under in-plane shear [63] or applied uniaxial stress [184, 279] due to the interaction with a substrate [172], due to internal or thermally induced stresses [19], due to thermal fluctuations [70], or under nanoindentation [86]. From atomistic continuum modeling it has been found that for freely suspended graphene the corrugations reach up to 1 nm over a lateral scale of 10-25 nm [170]. Understanding the different structural configurations of graphene, especially during deformation, is of great importance.

Figure 2.3 presents the stability region of flat graphene in the space of the in-plane strain components $(\varepsilon_{xx}, \varepsilon_{yy}, \varepsilon_{xy})$ as a set of sections by the planes $\varepsilon_{xy} = \text{const}$ projected on the $(\varepsilon_{xx}, \varepsilon_{yy})$ plane. Ripples cannot

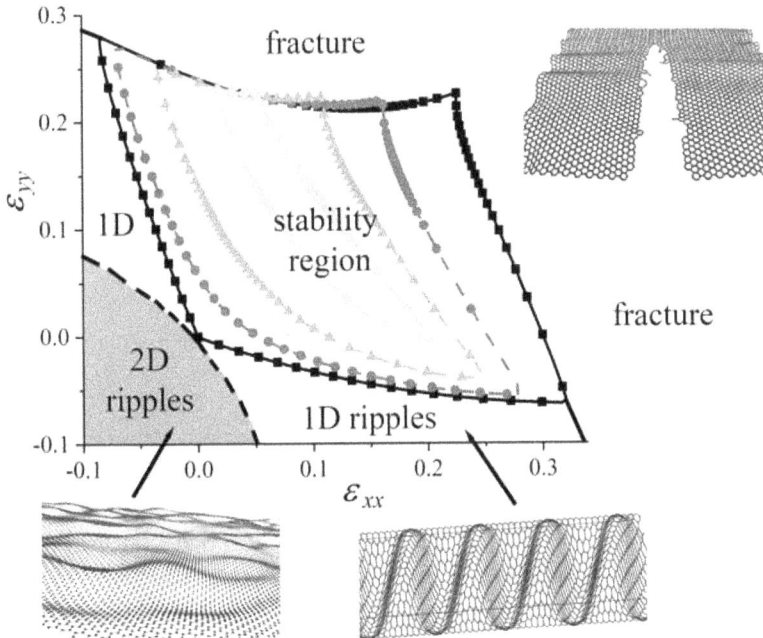

Fig. 2.3. (a) The stability region of flat graphene in the space of the in-plane strain components $(\varepsilon_{xx}, \varepsilon_{yy}, \varepsilon_{xy})$ shown as the set of sections by the planes $\varepsilon_{xy} = 0$ (black line), $\varepsilon_{xy} = 0.1$ (dark gray), $\varepsilon_{xy} = 0.2$ (silver) and $\varepsilon_{xy} = 0.3$ (light gray). Insets show the structural state of graphene in the different deformation regions.

form inside the stability region of flat graphene. It can be seen that with the increase of shear strain the stability region decreased, especially for $\varepsilon_{xy} > 0.2$. From the insets in Fig. 2.3 different structural states of graphene at different strain values can be seen: depending on the strain values two types of ripples can be found - 2D chaotic and 1D symmetric ripples. In the stability region, graphene remains flat.

Graphene can also be considered as the basic structure for a lot of other carbon polymorphs: CNTs can be obtained by graphene wrapping; fullerenes can be obtained by generating defects in graphene; diamane can be obtained from graphene layers under pressure, as well as diamond itself; graphite is simply a stacking of numerous graphene layers (greater than 100 layers); graphydiyne can be obtained by the adding of acetylenic bonds. That is why Geim and Novoselov called graphene "the mother of all graphites". Moreover, novel structures can be obtained from graphene – new composites, graphene aerogels, 2D van der Waals heterostructures, etc.

2.1.1.2 *Ripples in Graphene*

Ripples can appear in graphene due to very different factors: presence of the substrate, external strain, appearance of defects, graphene functionalization, but in all cases the main reason is the uncompensated stresses in the structure. The type, shape, amplitude and wavelength of ripples will be different for different stress states of graphene. The condition for the appearance of unidirectional ripples in in-plane strained graphene is that one of the principal membrane forces should be negative and another positive. Figure 2.4 presents the regions with different combinations of signs of principal membrane forces as the projection on the $(\varepsilon_{xx}, \varepsilon_{yy})$ plane for different shear strain: (a) $\varepsilon_{xy} = 0$, (b) $\varepsilon_{xy} = 0.1$, (c) $\varepsilon_{xy} = 0.2$, and (d) $\varepsilon_{xy} = 0.3$.

Region 1, where both membrane forces are greater than zero, is the stability region of flat graphene (the same as in Fig. 2.3). In region 3 where both membrane forces are lower than zero, ripples are non-unidirectional and chaotic. In this region corrugation of graphene occurred. In the white region, one of the principal membrane forces is positive and another negative, which results in the generation of stable unidirectional ripples.

The main parameters of 1D ripples are the orientation angle, the amplitude A, and the wavelength λ. All these parameters depend on the value and type of the applied strain. Figure 2.5a shows the expected orientations of the ripples for shear strain $\varepsilon_{xy} = 0.1$. Again, ripples can only appear in the region colored white, where the line segments oriented

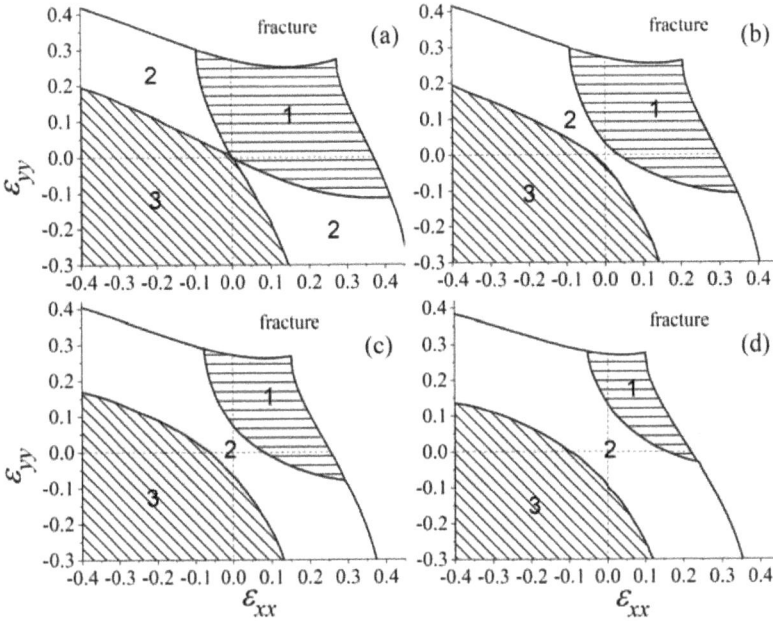

Fig. 2.4. Regions of the $(\varepsilon_{xx}, \varepsilon_{yy})$ plane with different signs of principal membrane forces for (a) $\varepsilon_{xy} = 0$, (b) $\varepsilon_{xy} = 0.1$, (c) $\varepsilon_{xy} = 0.2$, and (d) $\varepsilon_{xy} = 0.3$.

along the direction of the tensile membrane forces are presented. Without shear strain, the orientation angle of ripples can take only two values: 0° and 90°. For shear strain, the orientation of the ripples changes from 0° to 90° with decreasing ε_{yy} or increasing ε_{xx}. For example, for strain components $\varepsilon_{xx} = \varepsilon_{yy} = 0$ and $\varepsilon_{xy} = 0.1$, the orientation angle is 45°. Therefore, one of the ways to change the orientation of the folds is to apply shear deformation.

Examples of graphene with ripples of different orientation are presented in Fig. 2.5. At point A, the orientation angle of the ripples is 60°, at point B 45° and at point C 15°. The amplitude of the folds increases with distance from the boundary of the stability region. The wavelength of the folds at points A, B and C is 44, 37 and 34 Å, respectively. It is shown that the wavelength of the folds decreases with an increase in the shear component of the deformation.

Another possible way to obtain ripples in graphene is functionalization, for example, by hydrogen. As a result of partial hydrogenation, the new hybrid structure can be obtained with mixed sp^2/sp^3 hybridization. An example is presented in Fig. 2.6, where two rates of hydrogenation

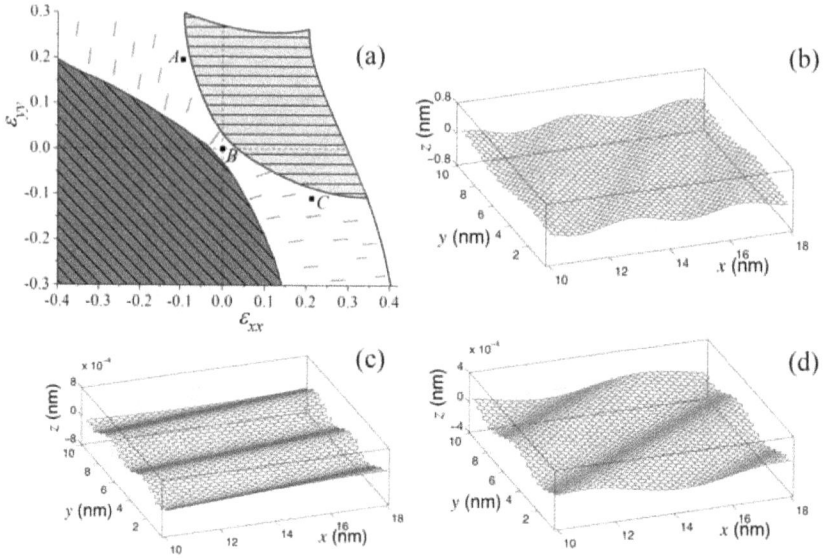

Fig. 2.5. (a) The stability region of flat graphene in the space of the in-plane strain components (ε_{xx}, ε_{yy}, ε_{xy}) shown for $\varepsilon_{xy} = 0.1$. Line segments show the orientation of ripples. (b–c) Examples of ripples at different strain: (b) point A from (a), (c) point C from (a), (d) point B from (a).

are considered, 25% (a-c) and 50% (a'-c'), with the formation of pure (sp^2) and hydrogenated (sp^3) stripes. Moreover, functionalization can be conducted in different ways: graphane (fully hydrogenated graphene), graphone (hydrogen atoms only on one side of graphene) and zigzag graphone. The name zigzag graphone was given to the structure because the hydrogen atoms are located within one zigzag line on one side of the graphene, while the next strip of hydrogen atoms is located on the other side of the graphene also within the zigzag line (see insets in Fig. 2.6).

MD simulation showed [153] that all hydrogenated regions deviated from the initial planar morphology, as shown in Fig. 2.6. It was found that sufficiently large periodic ripples are generated for all types of partially hydrogenated structures. The wavelength and amplitude of ripples are considerably dependent on the rate and type of hydrogenation. Thus, the parameters of ripples can be controlled by chemical functionalization.

The generation of ripples during hydrogenation is explained by the mismatch between the graphene lattice and the hydrogenated region, which results in the exceptional stresses. The lattice parameter for hydrogenated graphene increases by about 5%. Therefore, hydrogenation leads

graphane

(a)

graphone

(b)

zigzag graphone

(c)

(a')

(b')

(c')

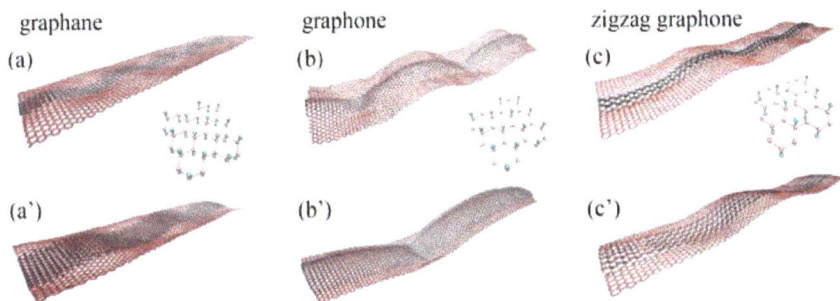

Fig. 2.6.   Ripples generated in GNR after hydrogenation. Distribution of hydrogen atoms over GNR is presented in the insets. Three different structures are considered – graphane, graphone and zigzag graphone. Two rates of hydrogenation of GNR, 25% (a-c) and 50% (a'-c'), are presented.

to the occurrence of stresses on the surface, leading to some graphene folding.

2.1.1.3   *Graphene Properties*

Interest in graphene is aroused by its large set of unique properties: high electrical and thermal conductivity, dependence of the electronic characteristics of graphene on the presence of attached radicals of various natures on the surface, quantum Hall effect, high mobility of charge carriers, transparency, unique elastic and deformation characteristics, etc., the most interesting of which are given in Fig. 2.7.

The phonon modes in graphene have been studied theoretically by DFT or molecular dynamics (MD) simulation for years since it is only possible to directly measure a very limited part of this dispersion relation using Raman spectroscopy. Analysis of the acoustic and optical phonons in graphene allow us to understand and characterize some of their properties. For example, the appearance of the low-frequency out-of-plane acoustic phonons results in negative thermal expansion coefficient of graphene. The Raman-active phonons in graphene allow the use of Raman spectroscopy for the measurement of, for example, the thickness of graphene.

The sound velocities of graphene can be calculated as:

$$V_{x,j} = \frac{\omega_j \Delta}{q_x p_{2,y} - q_y p_{1,y}}, \quad V_{y,j} = \frac{\omega_j \Delta}{-q_x p_{2,x} + q_y p_{1,x}},$$

where ω_j should be calculated for a small wavevector $|q| \leq \pi$; and $\mathbf{p}_1$ and $\mathbf{p}_2$ are strained lattice translation vectors, $\Delta = p_{1x} p_{2y} - p_{1y} p_{2x}$.

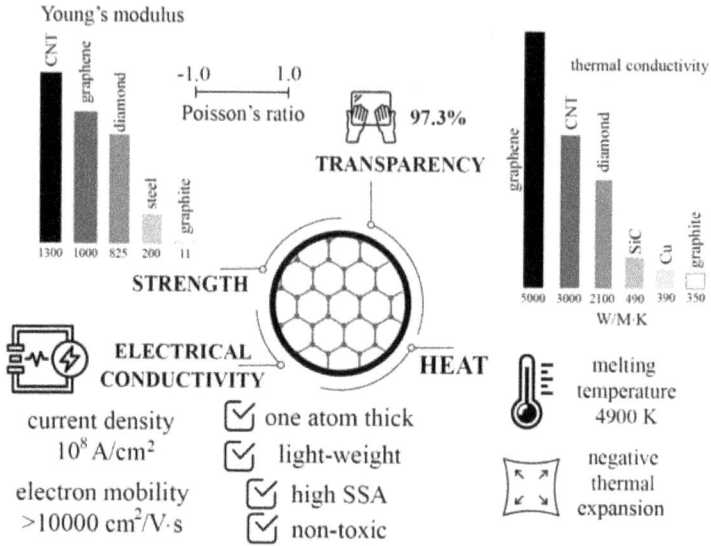

Fig. 2.7.   Exciting graphene properties.

Direction of vector **q** defines the propagation direction of sound waves.

Figure 2.8a presents the sound velocities of unstrained graphene. The phonon spectrum of graphene includes three acoustic and three optical branches. The acoustic branches with the highest (LA) and average (TA) frequencies correspond to longitudinal and transverse waves in the graphene plane, respectively. The acoustic branch with the lowest frequency (ZA) corresponds to transverse waves outside the graphene plane. In the unstrained state, graphene is isotropic and, therefore, the sound velocities do not depend on the propagation direction and are equal to 20.3 km/s for the LA phonon branch and 10.7 km/s for the TA phonon branch (from the MD calculation with Brenner potential). The ZA wave has zero sound velocity, since the bending rigidity of unstrained graphene is zero. Experimental works on the X-ray scattering show that the sound velocity in graphene is 22 km/s for longitudinal and 14 km/s for transverse waves [175], and the sound velocity for longitudinal waves obtained experimentally from Raman scattering is 20 km/s [90].

Figure 2.8b presents the density of the phonon states (DOS) of unstrained graphene obtained from MD. As can be seen, there is no band gap for infinite defect-free unstrained graphene.

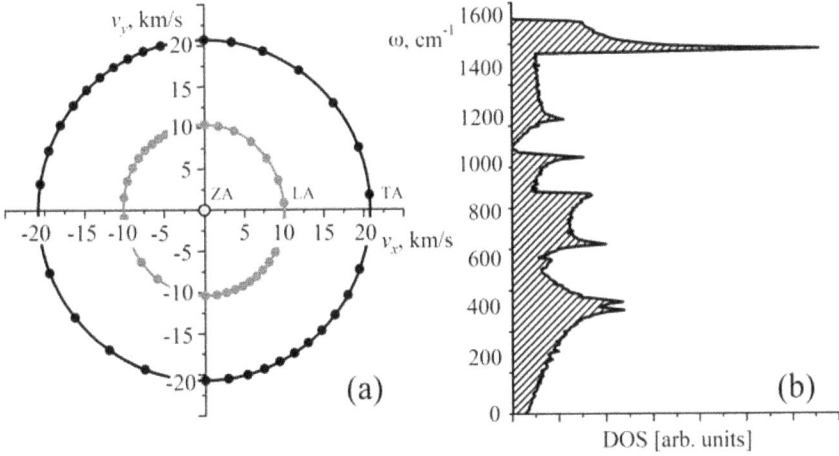

Fig. 2.8. (a) Sound velocities and (b) density of the phonon states (DOS) of unstrained graphene obtained from MD.

The electron mobility in graphene is more than 105 $\text{cm}^2\text{V}^{-1}\text{s}^{-1}$ at 300 K, which significantly exceeds the same parameter for silicon (Si) (1500 $\text{cm}^2\text{V}^{-1}\text{s}^{-1}$) or semiconductors such as AlGaAs/InGaAs (8500 $\text{cm}^2\text{V}^{-1}\text{s}^{-1}$) [10, 29, 61]. The maximum frequencies of the optical phonon spectrum (1600 cm^{-1}, see Fig. 2.8b) are also among the highest for known materials [84]. As a result, the thermal conductivity of graphene is comparable to that in diamond, and one order of magnitude higher than that of conventional semiconductors such as silicon.

The electronic properties of graphene are extremely attractive for practical application. It is known that electrons in graphene behave as relativistic particles that do not have mass (Dirac fermions). At low energies and long wavelengths, electrons in graphene are characterized not by their mass but by their propagation speed, the so-called Fermi-Dirac velocity, which is about 106 ms (about 300 times slower than the speed of light) [39]. At the same time, electrons in graphene obey relativistic wave equations and propagate in two dimensions.

The optical properties of graphene are also of great interest. It has been shown in the literature that single-layer graphene absorbs only 2.3% of incident light [10]. The combination of unusual electronic and optical properties opens up possibilities for the application of graphene in optoelectronics, photonics and spintronics.

It has been experimentally shown that the thermal conductivity of graphene is 5000 W/m·K. Generally speaking, as shown by various experiments, the thermal conductivity of graphene varies within wide limits depending on the experimental conditions, the graphene structure, etc. Changes obtained by Raman spectroscopy show a thermal conductivity coefficient from 1800 W/m·K to 5000 W/m·K [17, 85]. In addition, the results obtained by the thermal-bridge method describe the dependence of the heat transfer coefficient on the cross-sectional area at low temperatures until the limit of ballistic conductivity of pure graphene is reached [84]. The thermal conductivity coefficient, calculated in the same work, was 190 W/m·K at room temperature. The differences in the values of thermal conductivity can be explained by the different sizes of the studied samples, the presence of defects in the crystalline structure, the imperfection of the measurement method, the presence of impurities on a substrate, etc. The thermal conductivity coefficient of graphene obtained by the CVD method varies from 2600 to 3100 W/m·K for samples with a diameter of 2.9 μm to 9.7 μm [36]. The authors explain this by the imperfection of the measurement method (Raman spectroscopy), as well as the presence of grain boundaries and defects in graphene obtained by the CVD method.

For GNR, the thermal conductivity coefficient strongly depends on their size, presence of defects, doping, shape, stress state, substrate, chirality, edge influence, presence of folds, etc. For example, the thermal conductivity coefficient of a zigzag-oriented nanoribbon is approximately 30% higher than that of an armchair-oriented nanoribbon, and this difference disappears for nanoribbons wider than 100 nm. The dependence of the thermal conductivity coefficient on the nanoribbon width is also unusual: for a zigzag nanoribbon, thermal conductivity increases and then decreases with increasing width, while for an armchair nanoribbon, thermal conductivity increases monotonically. As studies have shown, the presence of folds in the structure of nanoribbons also has a significant effect, namely, it reduces the thermal conductivity coefficient. Another interesting case is graphene, in which the number of layers varies from two to five. As the number of layers increases, the heat transfer coefficient decreases, approaching the heat transfer coefficient of 3D graphite in magnitude.

2.1.1.4 *Defects in Graphene*

Graphene properties can be considerably affected by imperfections in the crystal lattice: vacancies, dislocations, dipoles, grain boundaries.

Defects can even change the topology or curvature of graphene. Figure 2.9 shows point defects characteristic for the graphene lattice which are vacancies and impurity (or substitution) atoms, Stone-Walles (SW) defect, dislocations and dislocation dipoles (DD). Impurities can be either in the form of substitution atoms or in the form of isotopic impurities, which mainly affect the phonon spectrum of graphene. Impurities in the form of adsorbed atoms mainly lead to local changes in the electric field.

The vacancy generation in the graphene lattice leads to the breaking of three short C-C bonds, which requires the energy of 7.8 eV. Due to the high value of nucleation energy, monovacancies (Fig. 2.9a) are unstable and can combine into divacancies (Fig. 2.9c) with nucleation energy of only 1 eV greater than the energy of a single vacancy and 6 eV less than the energy of two separated single vacancies. Vacancy clusters also can easily appear in graphene [126]. Figure 2.9d shows a rounded vacancy disk, but it is believed that there is a second type of vacancy disks of an elongated shape. Interestingly, the type of vacancy cluster does not have a significant effect on the results, while both types of defects greatly reduce the strength of carbon structures.

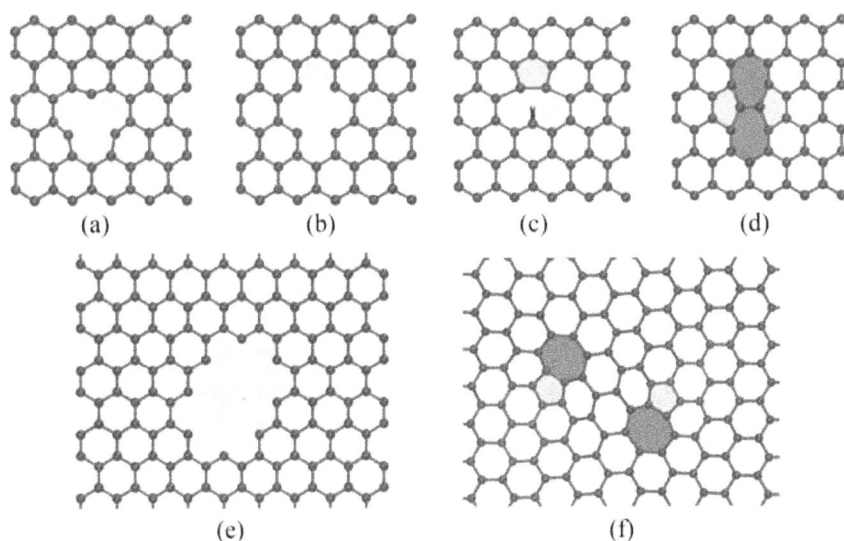

Fig. 2.9. Defects in the graphene crystal lattice: (a) vacancy, (b) impurity atom, (c) divacancy, (d) Stone-Walles defect, (e) vacancy cluster, and (f) dislocation dipole.

Another type of defect is penta- and heptagons, which are also called 5- and 7-atom ring defects or disclinations. Disclinations are point defects associated with rotational symmetry and are obtained by removing (adding) a wedge cut at an angle of 60° and then bringing the edges together. This procedure preserves the hexagonal lattice everywhere except for one unit cell, which becomes a pentagon (or heptagon). From Fig. 2.10 (see for details Ref. [158]) it is evident that the energy of defect 5 exceeds the energy of defect 7, and the total energy of defects 5–7 is the lowest, which explains the widespread occurrence of defects 5–7, or dislocations, in graphene. In order to obtain defect 5 geometrically, one can cut out a section of the structure and bring the edges of the cut together so that a cone is formed as a result. Defect 7 is obtained in a similar way: make a cut, spread the edges and insert a corner section. In this case, defect 5 can be considered a positive disclination, and defect 7 a negative disclination.

Heptagon is the same disclination as pentagon, but of the opposite sign; however, in this case, the axial symmetry is broken, since the graphene is arbitrarily curved (Fig. 2.10), which complicates the calculation process. It can be assumed that the curvature of such a structure decreases as $1/R$,

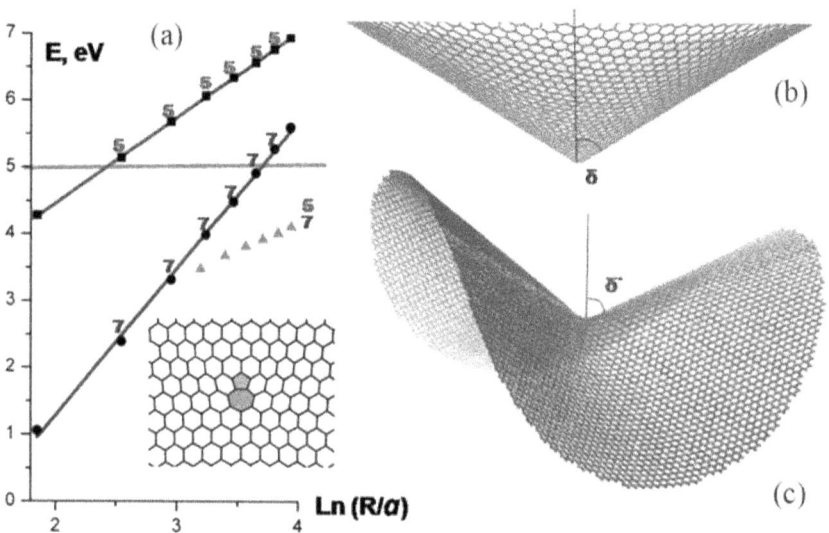

Fig. 2.10. (a) Defect energies computed for isolated 5 and 7 defects, as well as a 5–7 dislocation, as a function of size R of the lattice cluster. (b) Fully relaxed lattice containing a 5-atom ring. (c) Graphene lattice containing a single 7-atom ring [158].

and the total energy should increase with increasing sample size as $\ln R$, similar to a cone. According to this assumption, we write the energy of these defects as $E = E_c + E_e \ln R/a$. The energy E_c can be both positive and negative. An important observation is that 5 and 7 defects cause the formation of a non-planar structure, which is the reason for the observation of numerous curvatures in the experimentally obtained defective graphene. Recall that fullerenes also contain disclinations (pentagons), the presence of which leads to the formation of a spherical structure. Since the total energy of dislocation (5-7 defect) is lower than that for separate 5 and 7 defects, the nucleation of dislocation happens more often; this defect can be easily observed in experiment.

Defects 5-8-5 and 5-7-5-7 (or SW defect) also can be found in graphene as well as different grain boundaries [128, 157]. Examples of different defects are presented in Figs. 2.11–2.12.

Defect dynamics in graphene can considerably affect their mechanical properties and conductivity. The motion of single dislocations and DDs in graphene at zero and finite temperatures has been studied by different simulation methods [31, 122] and in experiment [261, 280]: dislocation annihilation, defect generation, self-healing of defects. The processes of defect generation, annealing, and rearrangement are determined by the activation energies (barriers) of thermally activated processes and the threshold energies of radiation-induced processes.

For DDs with different arm length the following movements were found: (1) dislocation gliding; (2) dislocation interaction with movement against each other; (3) generation of a SW defect when two dislocations approach each other; (4) annihilation; (5) dislocation climbing. For dislocation gliding

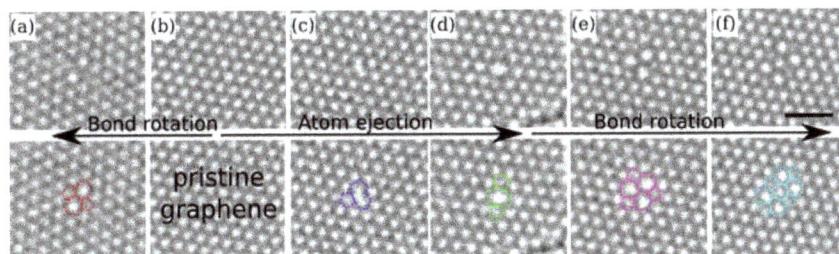

Fig. 2.11. Elementary defects and frequently observed defect transformations under irradiation. (a) Stone-Wales defect, (b) defect-free graphene, (c) 5-9 single vacancy, (d) 5-8-5 divacancy, (e) 555-777 divacancy, (f) 5555-6-7777 divacancy. Scale bar is 1 nm. Reprinted with permission from [128].

Fig. 2.12. Example of the symmetric tilt grain boundary structures in graphene. Reprinted with permission from [157].

for the DDs less energy is required for glide (5 to 10 eV) than for dislocation climb (9 to 12 eV). Figure 2.13 presents different DDs. The abbreviation of the dipole DD_k is introduced, where k denotes the distance between dislocations in hexagons: DD_0 – zero arm or SW defect, for DD_2 arm is 7 Å, for DD_3 arm is 9.3 Å, for DD_4 arm is 11.5 Å, for DD_6 arm is 16.4 Å, for DD_8 arm is 22 Å, and for DD_{10} arm is 30 Å.

For SW defect, only dislocation annihilation occurred when the C-C bond is rotated by 90°. The energy of SW defect is the lowest among all other defects in the graphene lattice. However, this defect is stable up to the temperature of 1400 K. As the temperature increases, the rotation of the C-C bond becomes easier, and both dislocation annihilation and the generation of a new SW defect in the lattice occur.

Different dynamic acts can be considered for dipole arm up to 16.4 Å, which is presented in Fig. 2.14.

The generation of new 5-8-5 defects can be observed additionally to the dislocation glide. The 5-8-5 defect is generated at sufficiently high temperatures ($T = 3400$ K) near one of the dislocations and can exist up to 30-50 ps. The 5-7 defect or SW defect, differently oriented, can also appear near one of the original dislocations. Such defect formation can be

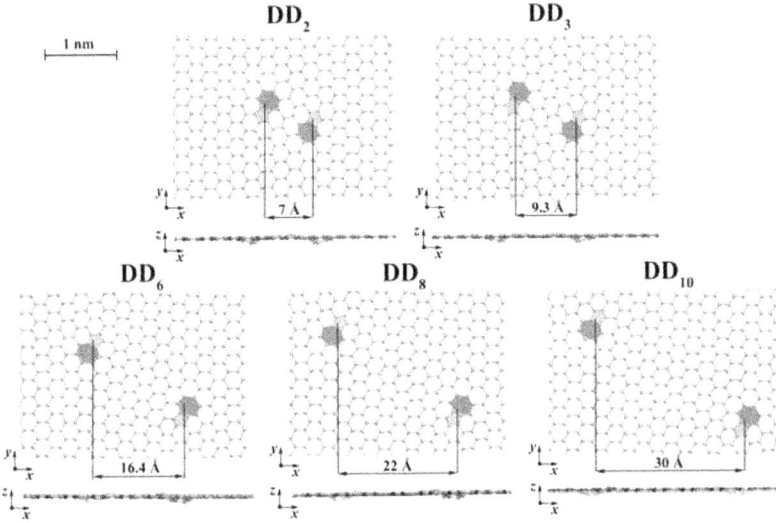

Fig. 2.13. Typical structure of graphene with DD. The five-atom rings are colored light grey, and the seven-atom rings are colored dark grey. All structures are presented as projections on the xy and xz planes. Reprinted with permission from [78].

the precursor for nucleation of grain boundaries. There is an interesting motion path when one dislocation glides toward another, and at the next time step, the other dislocation glides in the opposite direction. In this case, the distance between the dislocations remains unchanged; however, it is not the motion of the entire DD, but just two different dynamic steps.

For dipole arm 7 Å, the glide of one of the dislocations or the formation of SW defect occurred. The first act of dislocation glide in this case occurs at 2700 K. The formation of two SW defects can also occur. If SW defect appears, dislocations can further easily annihilate, since the temperature is much higher than 1400 K, at which annihilation is observed.

The longer the dipole arm, the less the possibility of defect annihilation. As the distance between two dislocations is less than 16 Å, continuous glide occurred. The first changes for different dipoles took place at different finite temperatures which can be seen from Fig. 2.15. Figure 2.15 presents the activation temperature T_A of different dynamic acts as the function of the dipole arm: filled circles are for first glide act, open circles are for second glide act, and squares are for SW generation.

For DDs with arms 9-11 Å similar dislocation dynamics is observed: (1) the glide of one of the dislocations, or glide of dislocations towards each other, (2) continuous glide of dislocations, (3) generation of SW defect

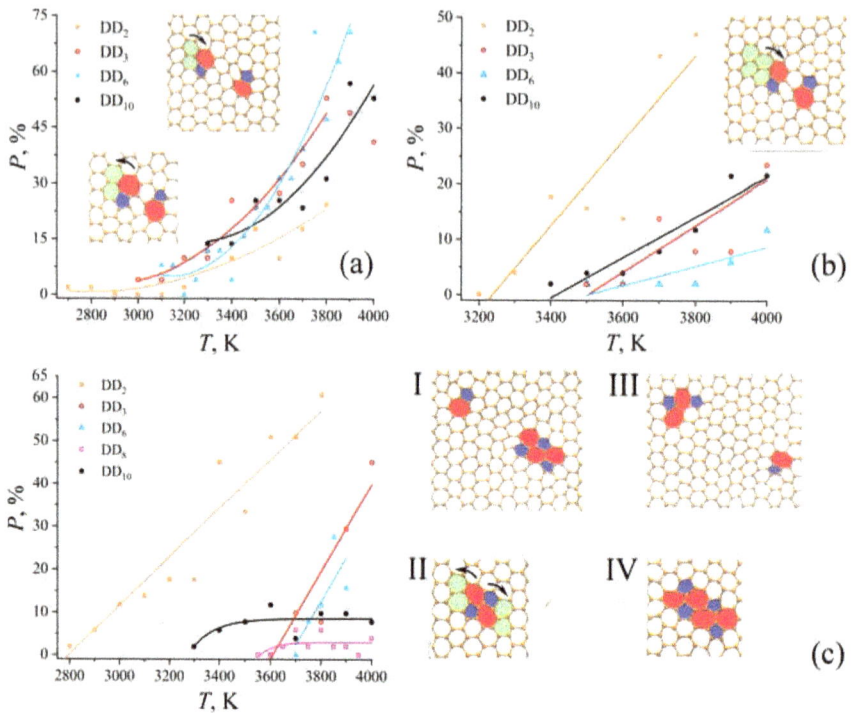

Fig. 2.14. The possibility of different dynamic acts in graphene with DD: (a) first gliding; (b) second gliding; (c) generation of the SW defect. The insets show the structural states for different dynamic acts. The abbreviation of the dipole DD_k is introduced, where k denotes the distance between dislocations in hexagons: DD_0 – zero arm or SW defect, for DD_2 arm is 7 Å, for DD_3 arm is 9.3 Å, for DD_4 arm is 11.5 Å, for DD_6 arm is 16.4 Å, for DD_8 arm is 22 Å, and for DD_{10} arm is 30 Å.

that can annihilate later and (4) dislocation glide in the direction opposite to another dislocation. The glide of dislocations in the opposite direction shows that for this dipole arm, dislocations can both attract and repel. Dislocation glide begins at 3000 K.

The DD with the arm length greater than 16 Å demonstrates a high stability and the motion of the dislocations is minimal: one or two glide events are observed for $T > 3000$ K. It is found that $l = 20$ Å is the critical value of the dipole arm at which the attraction and repulsion of dislocations occurred with the same probability. A further increase in the arm length reduces the probability of mutual motion of dislocations and increases the possibility of the formation of a static configuration.

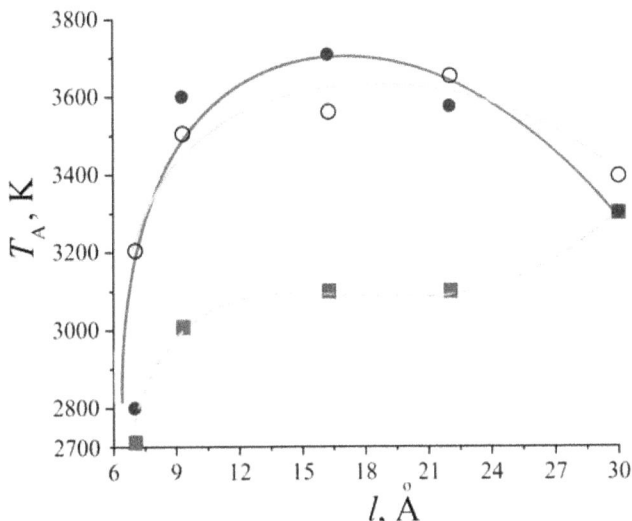

Fig. 2.15. Activation temperature T_A of different dynamic acts as the function of dipole arm: filled circles are for first glide act, open circles are for second glide act, and squares are for SW generation.

2.1.2 Carbon Nanotubes

CNTs are extended structures in the form of a hollow cylinder, consisting of one or several graphite layers rolled into a tube with a hexagonal organization of carbon atoms. The diameter of CNTs varies from one to several tens of nanometers, and the length is measured in tens of microns and is constantly increasing as the technology for their production improves.

The indices (m, n) are determined by the structure of the hexagonal grid cut out of the graphene sheet before rolling; m and n are the coefficients of the expansion in basis vectors of the segment connecting two atoms on opposite sides of the tape. When rolling, these atoms overlap. If, when rolling the graphene tape, each pair of sides of the hexagons is located at an angle to the tube axis different from 0 or 90°, then such nanotubes are called chiral. If two sides of each hexagon are perpendicular to the tube axis, then these are armchair. In zigzag tubes, two sides of each hexagon are parallel to the tube axis. The chiral angle characterizes the deviation from the zigzag configuration, varies from 0 to 30° and is calculated as

$$\theta = arctg[-\sqrt{3m}/(2n + m)],$$

or

$$\theta = arctg[-\sqrt{3n}/(2m + n)].$$

CNTs $(m, 0)$ are called zigzag nanotubes, and (m, m) are called armchair nanotubes (see Fig. 2.16). Nanotubes can be open or closed at one or both ends. In closed nanotubes, the ends of the tubes end in hemispherical caps made up of hexagons and pentagons, resembling the structure of half a fullerene molecule. The presence of caps at the ends of CNTs allows us to consider nanotubes as a limiting case of fullerene molecules, the length of the longitudinal axis of which significantly exceeds the diameter. The properties of such tubes can be controlled to a certain extent by changing their chirality, i.e., the direction of twisting of their lattice relative to the longitudinal axis. CNTs are obtained both with a metallic type of conductivity and with a given bandgap.

Multi-walled nanotubes can be composed from several to tens of single-walled nanotubes. Figure 2.17 presents the atomically resolved high-resolution transmission electron microscope (HRTEM) images of the individual B- and N-doped single-walled CNTs in the bundles. Corresponding computer-simulated HRTEM single-walled CNT images and structural models are shown in the insets [254]. The distance between the walls approaches the interlayer distance in graphite (0.34 nm). For this reason,

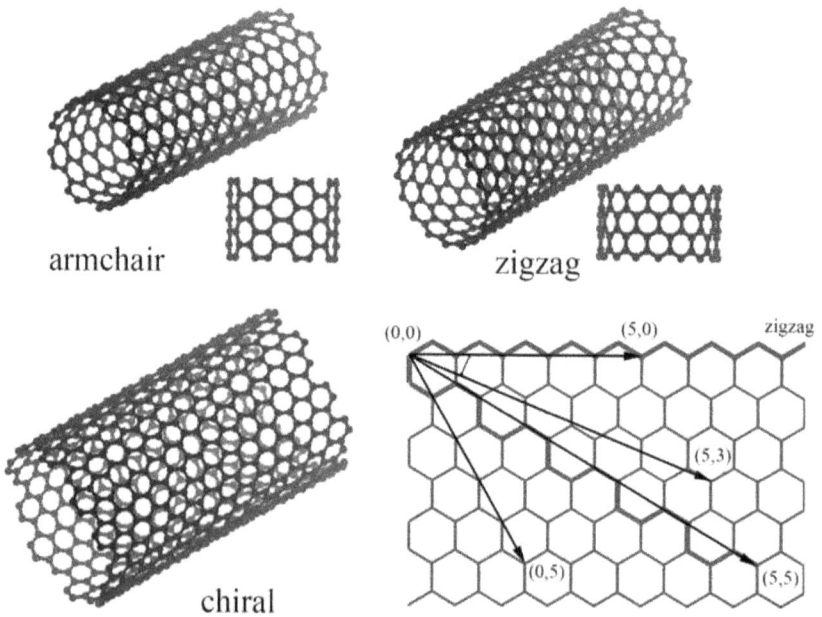

Fig. 2.16.   Single-walled CNTs: armchair, zigzag and chiral.

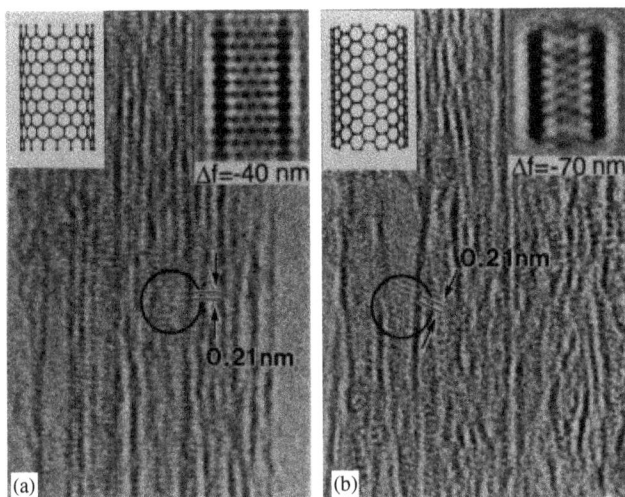

Fig. 2.17. Atomically resolved HRTEM images of the individual B- and N-doped single-walled CNTs in the bundles: (a) A zigzag single-walled CNT; (b) an armchair single-walled CNT. Reprinted with permission from [254].

the minimum diameter of a single-layer nanotube is 0.7 nm, and subsequent diameters are determined by this minimum value. As a result, the inner and outer diameters of a multi-layer nanotube have values of 0.7-4 nm and 5-40 nm, respectively.

The chirality indices of a nanotube (m, n) uniquely determine its structure, in particular its diameter D. This connection is obvious and has the following form:

$$D = \frac{\sqrt{3}d_0}{\pi}\sqrt{m^2 + n^2 + mn}, \qquad (2.1)$$

where $d_0 = 0.142$ nm is the distance between adjacent carbon atoms in the graphite plane.

Double-walled CNTs can consist of two coaxial graphene monolayers with different structural and electrical characteristics. In double-walled CNTs, the outer tube protects the inner tube from chemical and mechanical influences, while the inner tube is a mechanical support for the outer tube (as in multi-walled nanotubes). The inner tube has a diameter comparable to single-walled nanotubes. Such CNTs are more resistant to mechanical influences than single-walled CNTs or fullerene-filled single-walled CNTs.

Double-walled CNTs are of great interest in studies of the evolution of nanostructure properties with an increase in the number of atomic layers. They provide a unique opportunity to study the interaction between concentric CNT walls.

To date, various methods to synthesize CNT nanostructures were developed [27, 28, 129, 188]. The most widely used method of CNT fabrication uses thermal spraying of a graphite electrode in an arc discharge plasma in a helium atmosphere. This method allows obtaining of CNTs in quantities sufficient for a detailed study of their physicochemical properties. In a direct current arc discharge with graphite electrodes at a voltage of 15–25 V, a current of several tens of amperes, and an interelectrode distance of several millimeters, intense thermal spraying of the anode material occurs. Spray products containing, along with graphite particles, also a certain amount of fullerenes are deposited on the cooled walls of the discharge chamber, as well as on the surface of the cathode, which is colder than the anode. CNTs were first discovered in this cathode deposit. In this case, the relative content of CNTs in the cathode deposit does not exceed several percent. Using an electric arc, mostly multi-layer nanotubes are synthesized, the diameter of which varies in the range from one to several tens of nanometers. In addition, such nanotubes are distinguished by chirality, which determines the difference in their electronic structure and electrical characteristics. The distribution of nanotubes by size and chirality angle critically depends on the arc conditions and is not reproduced from one experiment to another.

CNTs can be obtained by thermal spraying of graphite. When synthesizing CNTs by thermal action of laser radiation, the use of metal catalysts leads to the same qualitative effect as in the case of electric arc synthesis discussed above. The content of nanotubes in soot increases sharply, and their quality improves significantly. In this case, instead of multi-layer CNTs, characterized by a significant difference in diameter, single-layer CNTs are synthesized, the diameter of which varies in a relatively small range.

CVD has been used since 1952: hydrocarbon gas (for example, acetylene) passing over an iron, nickel or cobalt catalyst at temperatures of 500–800°C. The growth mechanism consists of the decomposition of a hydrocarbon molecule on a catalyst particle (5–20 nm in size) with subsequent formation of CNTs. The diameter of the CNT is approximately equal to the size of the catalyst particle. Reactor also contains a gas environment (nitrogen, hydrogen, argon). This type of reactor is one of the most often used for the mass production of CNTs. The final product contains multi-walled CNTs, the defectiveness of which is lower than with an arc discharge. Of great interest is the electrolytic synthesis of CNTs. At the same time, CNTs of various morphologies, as well as other carbon phases, such as spiral or rounded structures, were found among the electrolysis products.

As with graphene, there are the same defects in CNTs: the most common are vacancies and SW defects. Vacancy-type defects are nucleated during processing after the synthesis of CNTs, for example, as a result of exposure to a high-energy electron or ion. When vacancies are formed, rehybridization of three released bonds or their binding to molecules from the space surrounding the nanotube is possible. The broken bonds near the vacancies – valence electron shells – can, when rearranged, serve as sites for bonding with other CNTs, provide chemical sensitivity, or attach impurities.

Divacancies can appear as a result of interaction of single vacancies with each other during thermally activated movement along the CNT. In graphene and CNTs, a double vacancy (divacancy) is formed by the loss of two carbon atoms. Due to the additional stress from the surface bending, divacancies have a formation energy 1.5 eV lower than monovacancies. During annealing, interaction of single vacancies and growth of large voids on CNTs is observed, similar to what happens for graphene. In single-walled CNTs, the barrier to vacancy movement is 1 eV, thereby ensuring the mobility of the latter at low (100-200°C) temperatures. A divacancy can be reconstructed into a pentagon-octagon-pentagon (5-8-5) defect.

An interstitial atom is an additional atom formed between the layers of a multi-walled CNTs. Interstitials have a relatively high formation energy (about 5.5 eV) and usually do not form in CNTs during synthesis (except for the laser ablation method and the discharge-arc method). When a multi-walled CNT is irradiated with high-energy particles, an atom of the crystal lattice is released, which usually remains within the multi-walled CNT and participates in the formation of a new bond between two adjacent layers. An interstitial atom is highly mobile in the interlayer space of a multi-walled CNT and relatively easily forms a vacancy-interstitial complex. For the same reasons, these defects have a short lifetime in CNTs. In addition to recombination with a vacancy, the formation of an agglomerate from several interstitial sites or the binding of an interstitial atom with adsorbates and subsequent assembly into small clusters of amorphous carbon is possible.

The appearance of vacancies leads to changes in the absorption spectra in the infrared range. As the analysis shows, these changes are associated primarily with the appearance of vibrational modes localized mainly near the region of the vacancy or defect. As a result of the appearance of a vacancy or defect, the absorption spectra in the ultraviolet range become richer. From general considerations it is clear that these changes are caused by a shift and splitting of energy levels, which is associated with a decrease in the symmetry of the nanotube caused by the appearance of a vacancy or defect. Note that another type of defect is the cap, which sometimes appears during synthesis at the end of the nanotube and can affect CNT properties.

The majority of properties are defined by the CNT type – armchair, zigzag or chiral. Even the first studies of the electrical properties of CNTs showed that they can be both metallic and semiconducting which depend on the geometry of the nanotubes. The nanotubes for which the difference $n - m$ is a multiple of three turned out to be metallic, and the rest are semiconductors. In the metallic state, the conductivity of the nanotube is very high. According to estimates, they can pass a current density of approximately 109 A/cm^2, while copper wire fails already at a current density of about 106 A/cm^2.

Important characteristics of the nanotube such as the band gap, electrical resistance, concentration and mobility of charge carriers are determined by its geometric parameters – diameter and chirality, i.e., the orientation angle of the graphite surface relative to the tube axis.

There are currently a huge number of experimental and theoretical works on the specific heat capacity and thermal conductivity of CNTs. It is generally recognized that the main contribution to the heat capacity of CNTs and their bundles is made by phonon modes even at low temperatures. To date, there are two approaches to the theoretical study of the vibrational heat capacity of single-walled CNTs: numerical modeling of the vibrational dynamics of CNTs, in particular, the force constant model, and the MD method. However, the results obtained may differ greatly. Modern ideas about the low-temperature heat capacity of single-walled CNTs are contradictory, which is associated with both the methods of nanotube synthesis and the calculation methods. In particular, it is practically impossible to estimate the heat capacity of an individual CNT in an experiment. In addition, defects, the presence of a substrate or external environment, the multi-layer nature of nanotubes, the purity of the structure or the accuracy of the theoretical calculation all have a great influence.

One of the reasons for the high conductivity of CNTs is the very small number of defects that cause electron scattering. This is also facilitated by the high thermal conductivity of nanotubes. It is almost twice as high as that of diamond. At $T \geq 1$ K, the conductivity is a linear function of the logarithm of the temperature and reaches saturation at lower temperatures.

Mechanical characteristics can have a great impact on the conductivity of CNTs. As a result of deformation, the width of the band gap, the carrier concentration, the phonon spectrum, etc., all change. Thus, bending a nanotube at an angle of 105° leads to a decrease in its conductivity by 100 times. This property of a nanotube can be used as a basis for a nanodevice such as a converter of a mechanical signal into an electrical one and vice versa. According to estimates, the resistance of CNTs is approximately 2-3 orders of magnitude lower than that of copper.

The main parameter characterizing the tensile strength of the CNT is the longitudinal Young's modulus. The Young's moduli of CNTs obtained to date usually vary within the range from 1 to 1.3 TPa. Value of the elastic modulus can be very different for single-walled and multi-walled CNTs. For example, the elastic moduli for multi-walled CNTs are 810±410 GPa. The maximum strength of single-walled CNTs varies from 13 to 52 GPa with moduli from 320 to 1470 GPa. At the same time, for multi-walled CNTs, the tensile strength for the outer CNT is 11–63 GPa, moduli 270–950 GPa, and elongation 12%. Note that the values of the elastic modulus obtained in different experimental and theoretical works can significantly differ from each other, which is defined by the experiment conditions and by the CNT structure.

In order to experimentally measure the Young's modulus of a CNT, special methods have been developed, for example, based on the analysis of thermal vibrations of the free edge of the CNT, which is fixed at the other end [130, 267, 293]. Considering a nanotube as a hollow cylinder with a given wall thickness, it is possible to derive a relationship between the amplitude of the oscillations of the edge of the cylinder and the modulus of elasticity. A sufficiently large number of single-walled CNTs was studied and the Young's moduli $E = 1.25 - 0.35 \pm 0.45$ TPa were obtained.

Another important property of CNTs is the specific surface area which reaches a record value of 2600 cm^2/g^{-1}. Due to high specific surface area and natural curvature of the CNT, it is capable of absorbing gaseous and liquid substances. Since the diameter of the internal channel of the CNT is only 2-3 times greater than the characteristic size of the molecule, the capillary properties of the nanotube are manifested on a nanometer scale.

2.1.3 *Fullerenes*

Fullerene (or buckyball) is a broad class of polyatomic carbon molecules with the general formula C_n (n is an even integer) that have the shape of a closed hollow polyhedron. As was fully described in the Nobel lecture of Harold Kroto [134], the appearance of C_{60} fullerenes was first revealed in 1985 from the mass spectrum presented in Fig. 2.18a. The best-known fullerenes are C_{60} and C_{70} (Fig. 2.18b).

In accordance with Euler's theorem, which determines the number of polygons for flat and closed surfaces, it follows that 12 pentagons and any number of hexagons are required to form a closed surface (except for 1 hexagon, because it is impossible to obtain C_{22} fullerene). The number of hexagons N_{hex} for each individual spherical molecule can be calculated as:

$$N_{hex} = \frac{N}{2} - 10,$$

where N is the number of atoms in the molecule.

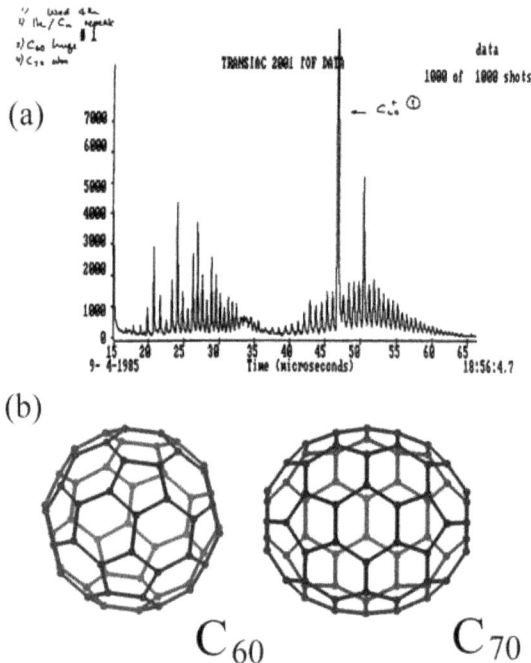

Fig. 2.18. (a) Mass-spectrum on which the dominance of C_{60} fullerene signal was first recorded. Reprinted with permission from [134]. (b) The most widespread fullerenes: C_{60} and C_{70}.

Fullerenes with no adjacent pentagons are the most stable: each pentagon is surrounded by 5 hexagons and has common edges only with hexagons. The structure of the majority of synthesized and isolated fullerenes obeys the rule of isolated pentagons.

The C_{60} molecule is the most symmetrical of all known so far and consists of 60 carbon atoms located on a spherical surface with a diameter of about 1 nm (Fig. 2.18b). Carbon atoms in fullerenes are located on the surface of the sphere at the vertices of pentagons and hexagons. The C_{60} molecule exactly repeats the shape of a soccer ball, having 12 black pentagons and 20 white hexagons. The double bonds are placed between hexagons and called (6,6), while others are placed between pentagons and hexagons and called (5,6). In accordance to [34], the length of double bonds is in the range of 1.369-1.406 Å, while for single bonds it is 1.43-1.467 Å. Fullerene C_{60} has internal diameter 0.5 nm and external diameter 0.71 nm. For fullerene C_{70} there are five hexagons with delocalised π bonds.

To date, different fullerene molecules were synthesized and studied, with the numbers of carbon atoms from 36 to 960 and more. Figure 2.19 presents the different types of fullerene molecules and a fullerene onion $C_{20}@C_{80}$. The smallest possible fullerene C_{20} consists entirely of pentagons. In a fullerene molecule, carbon atoms are linked to each other by a covalent bond. In fact, a fullerene molecule is an organic molecule, and the fullerene itself is a molecular crystal.

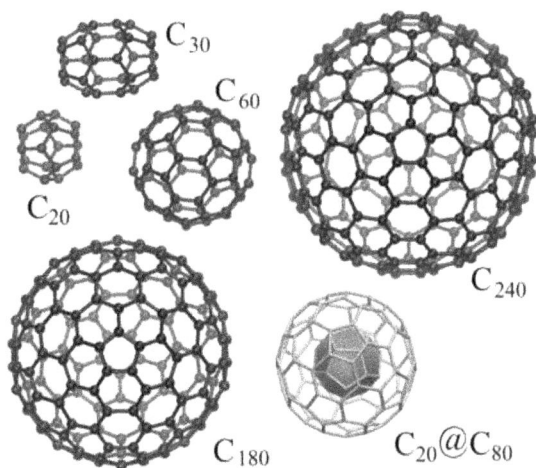

Fig. 2.19. Different possible fullerenes and onion $C_{20}@C_{80}$.

Nevertheless, today there are both theoretical and experimental works confirming the possibility of the existence of various fullerene derivatives (endohedral and exohedral). Metal ions can be used to functionalize fullerene (i.e., change its properties), forming a fundamentally different system. For example, alkaline elements (Na, K, Rb, Cs) are standard electron donors, and most of the work on doping fullerenes is devoted to the use of these metals. Also common methods for improving the properties of fullerenes are the addition of substitution atoms or the entrapment of an atom of another element inside the fullerene.

The process of synthesizing crystalline fullerene turned out to be quite simple. In an arc discharge using graphite electrodes in a helium atmosphere, soot is formed, which is then dissolved in benzene or toluene. From the resulting solution, gram quantities of C_{60} and C_{70} molecules in a ratio of 3:1 and 2% of heavier fullerenes are isolated in pure form. Currently, C_{60} fullerenes are a fairly accessible and widely used material. Now, there are many ways to obtain fullerenes: heating graphite rods with electric current in a vacuum, electric arc discharge between graphite electrodes in a helium atmosphere, and burning carbons, hydrogens and naphthalene. As a result of synthesis, a complex mixture is formed containing carbon soot, a mixture of fullerenes of various compositions and impurity molecules, usually of polyaromatic composition. Fullerenes are isolated by extraction with organic solvents, followed by separation into individual products; however, methods of synthesis, separation and purification are constantly being improved.

The very high hardness of fullerenes allows them to be used to produce fullerite micro- and nanotools for processing and testing superhard materials, including diamonds. For example, fullerite pyramids made of C_{60} are used in atomic force probe microscopes to measure the hardness of diamonds and diamond films. Fullerenes are also widely studied as materials for electro-optical studies. Fullerenes and compounds based on them are promising materials to obtain new nanostructures with improved properties. For example, fullerene films can be used to create 2D photonic crystals, and the optical properties of fullerene films can be changed by adding semiconductor materials, such as CdSe and CdTe. At room temperature, pure fullerene is an insulator with a band gap of more than 2 eV or an intrinsic semiconductor with very low conductivity.

The high binding energy of carbon atoms in fullerene molecules and the symmetry properties of these molecules determine their anomalously high thermal stability. Fullerene C_{60} loses its chemical structure only when heated to a temperature above 3000 K. The study of the C nanoclusters with number of atoms larger than 30 shows that the stability of clusters with even atomic number significantly exceeds the stability of clusters with odd atomic number. Moreover, fullerenes are also very resistant to collisions. For example, the collision of charged fullerenes C_{60}, C_{70} and C_{48} with the surfaces of purified graphite and silicon at an energy varying in the range up to 200 eV leads to the loss of a significant part of the kinetic energy, but is not accompanied by fullerene decomposition.

Fullerene exhibits unusual behavior in solutions. For example, fullerene C_{60} is practically insoluble in polar solvents such as alcohols, acetone, tetrahydrofuran, etc. It is slightly soluble in alkanes such as pentane, hexane, and decane, and with an increase in the number of carbon atoms, the solubility in alkanes increases.

2.1.4 *Graphane*

Graphene has an extremely large specific surface area (SSA) and, in combination with its light weight, strength, and unique physical and chemical properties, is one of the materials most suitable for hydrogen storage and transportation. Although graphite is known as one of the most chemically inert materials, single-layer graphene can interact with atomic hydrogen, which turns this highly conductive semimetal into a semiconductor. Covalent attachment of chemical groups on graphene leads to a change in sp^2 hybridization to sp^3 hybridization and the opening of a gap in the graphene spectrum. Moreover, chemical modification can change the optical, chemical and mechanical properties of graphene.

It is known that when atoms of other chemical elements are deposited on graphene, it loses its flat shape: the deposited atoms pull the carbon atoms out of the plane. The structure of graphane, fully hydrogenated graphene with H atoms alternatingly above and below the layer, is presented in Fig. 2.20 in two projections and in a perspective view. It was first predicted theoretically in 2007 [260] and then obtained experimentally in 2009 [66].

Double-sided hydrogenated graphene is considered as graphane, although commonly graphene is just partially hydrogenated. Moreover, hydrogenation can be conducted in a different manner: Figure 2.20c shows different types of hydrogen distribution over the surface of graphene – chair, stirrup, twist-boat, boat-1, boat-2, and tricycle configurations [71, 305]. Note that instead of hydrogen, fluorine atoms can cover the graphene

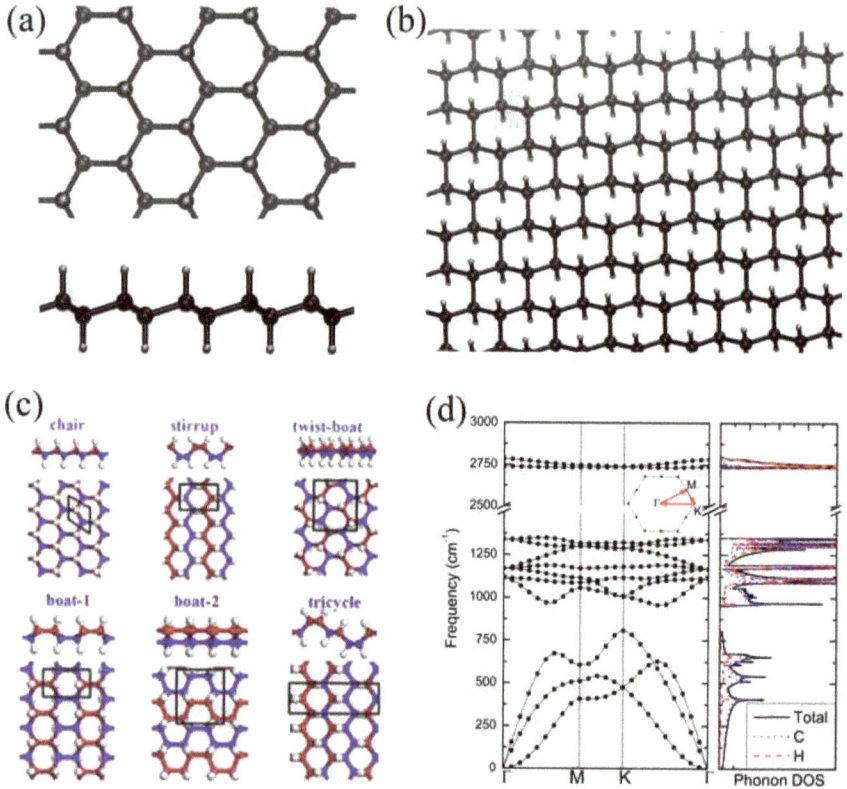

Fig. 2.20. (a,b) Graphane – fully hydroginated graphene: two projections and a perspective view. (c) Side and top views of graphane crystal structure with chair, stirrup, twist-boat, boat-1, boat-2, and tricycle configurations. The red and blue balls correspond to carbon atoms with up and down hydrogenation, respectively, and the white balls are hydrogen atoms. Reprinted with permission from [305]. (d) Phonon dispersion of graphane in the chair conformation. The dots are the directly calculated frequencies, and the lines are interpolated values. The inset shows the first Brillouin zone and the wavevector path used. The right-hand side shows the phonon DOS, where the black lines indicate the total DOS, the blue dotted lines the C projected DOS, and the red dashed lines the H projected DOS. Reprinted with permission from [201].

and the structure will be called fluorographene. It demonstrates different mechanical properties, heat resistance and has an acceptable band gap.

Figure 2.20d presents the phonon spectrum of graphane [201]. The obtained spectrum exhibits a wide gap between 1120 and 2750 cm^{-1}. The high-frequency optical phonon gap (above 2500 cm^{-1}) is generated by the hydrogen vibrational modes, while the low-frequency acoustic phonon branch (below 500 cm^{-1}) is determined mainly by the carbon vibrational modes. The C and H modes can also be divided into modes along the x or y directions (xy-modes) and along the z direction (z-modes). The in-plane modes (xy-modes) do not have a frequency above 1250 cm^{-1}, i.e., the high-frequency modes are out-of-plane modes (z-modes).

Attachment of the hydrogen atoms results in the breaking of the continuity of the π-bond, decrease of electron mobility and increase of electronic band gap. No significant difference in band gaps was observed for different types of hydrogenated graphene, but there is considerable dependence on the hydrogenation degree. Changing the hydrogen coverage, one can change the gap to 0 eV as for graphene. Thus, hydrogenated graphene possesses a tunable band gap which ranges from 0 to 5.4 eV, endowing its semiconducting properties and low electrical conductivity. Even one H atom per 100 000 carbon atoms reduces the electron transfer more than five times [71].

In contrast to graphene's very small spin–orbit coupling, graphane has enhanced spin–orbit interaction and local magnetic moments which result in ferromagnetism. Hydrogenated graphene exhibits little light absorption for light irradiation energy lower than 3.8 eV, and possesses photoluminescent properties, which is promising for applications in optical and optoelectronic devices.

Since hydrogen deposition on the graphene surface is a reversible process and can be carried out in a controllable manner, it is possible to create hybrid samples of hydrogenated graphene that will contain fully or partially hydrogenated stripes. In this case, the transformation of sp^2 hybridized graphene atoms into sp^3 atoms during hydrogen adsorption occurred partially and the structure will have mixed hybridization and different properties. As a result, graphene can be divided into parts with different thermal, electrical or magnetic properties. This means that insulating/antiferromagnetic or conducting/ferromagnetic properties can be obtained for the same graphene sample, depending on the surface functionalization. Thus, graphene can be used to obtain nanoelectronic devices by simple hydrogenation.

Hydrogenated graphene can be considered as a prospective structure for hydrogen storage and transportation. Hydrogen can interact with the graphene surface via van der Waals forces or by covalent chemical bonding with carbon atoms. The efficiency of hydrogen storage is usually estimated by two parameters: gravimetric (weight of stored hydrogen relative to the weight of the system in percent) or volumetric (accumulated mass of hydrogen per unit volume of the system) density. The interaction between hydrogen and graphene can be controlled by changing the distance between adjacent layers or by chemical functionalization of the material, which is used to enhance the adsorption/desorption properties of hydrogen on graphene. The binding of molecular hydrogen is weak and therefore requires low temperatures at high pressures to achieve stability of its accumulation [211]. Under high pressure and low temperature, H_2 can uniformly cover the graphene surface. Chemical adhesion of atomic hydrogen is an energetically more favorable process, since the H–H bond energy is slightly higher than the C–H bond energy. The formation of hydrogen dimers on the graphene surface requires more energy than isolated hydrogen.

2.1.5 *Diamane and Diamond-Like Phases*

One of the new 2D structures which can be directly obtained from graphene is diamane, which was theoretically proposed in [48]. Diamane is composed of two graphene layers connected by a covalent bond and covered with hydrogen atoms from both sides (see Fig. 2.21) [92, 120, 144]. Diamane is a 2D diamond-like structure (the thinnest diamond film) with the dangling carbon bonds from both sides saturated with hydrogen (H), fluorine (F) or chlorine (Cl) atoms. In literature diamane can also be

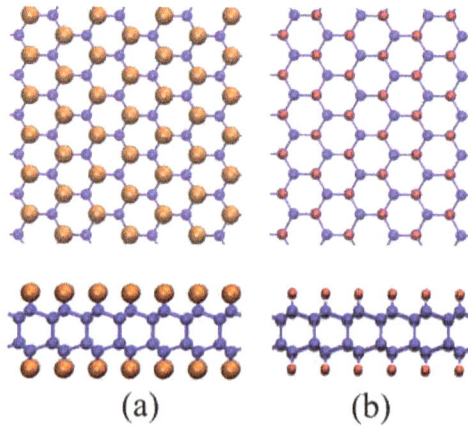

(a)　　　　(b)

Fig. 2.21. Diamane with (a) Cl and (b) H fictionalization in two projections. Carbon atoms are shown in blue.

known like bilayer graphane (hydrogenated or fluoridated) [146], bilayer diamond [219], and diamond-like/diamondized bilayer graphene [107, 197]. The surface functional groups can change the band structure, optical and mechanical properties of diamane [81, 177, 206, 250, 253, 274]. Due to the stronger electronegativity of F atoms compared with H atoms, F atoms are more easily functionalized [154], and fluorinated bilayer graphene has a shorter growth time under CVD in contrast with other functionalization methods [287].

At first, the binding of graphene layers was achieved by applying high pressures [80]. Diamane could be obtained from bilayer graphene with the presence of H or F atoms, which has been predicted in theory [242] and demonstrated in experiments [68, 207, 225]. The fluorinated diamane (F-diamane) was synthesized from bilayer graphene on CuNi(111) substrate [16]. After the transformation of bilayer graphene to diamane it still can be unstable, which can be overcome by chemical surface functionalization (such as hydrogenation and fluorination). Nowadays, fabrication of diamane with size up to several centimeters can be achieved [208].

Figure 2.21 presents the diamane structure in two projections. Diamane has two stable configurations [47, 48, 141]: AA is composed of graphene layers located one above the other, and hydrogen atoms are arranged in a checkerboard pattern; and AB is a Bernal-stacked form of bilayer graphene. Both structures have hexagonal symmetry. For convenience, further diamanes will be abbreviated as D-AA and D-AB. Moreover, since structures with and without hydrogen are considered, the abbreviations D-AA+H and D-AB+H will be further used. The thickness of diamane is equal to 6.8 Å for structures with hydrogen and 4.6 Å without hydrogen [223, 285].

Diamanes have a high bending stiffness of 3600 eV·Å [285] and wide band gaps in 1.2 to 2.8 eV range [46], and exhibit a semiconductor-like electronic structure, characterized by a direct bandgap [146]. However, diamane has a forbidden bandwidth in the range of 2.527–4.153 eV [177, 242]. The great advantage of this structure is that the bandgap of diamane is tunable concerning thickness, functional group type, and conformation [81, 180–182]. Janus diamanes with varying halogens can produce excitons with binding energies exceeding 1 eV [249]. The thermal conductivity of different diamanes at 300 K are 1960 $Wm^{-1}K^{-1}$ and 2240 $Wm^{-1}K^{-1}$, respectively [310].

2.1.6 *Peapods and Onions*

The combination of different carbon polymorphs into new low-dimensional structures, sometimes called all-carbon compounds, is also of high interest nowadays, because they demonstrate some unique and interesting properties.

An important class of all-carbon structures are onions. Like multi-layer CNTs, such structures are multi-layer fullerenes. Since the sizes of fullerenes vary from the smallest C_{20} with a radius of 3.68 nm to the largest C_{960} with a radius of 27 nm, it is possible to combine fullerenes into layered structures in which some fullerenes are enclosed inside a lattice of larger fullerenes. In an equilibrium structure, the distance between the fullerene layers should be 0.335 nm, which is very close to the equilibrium interplanar distance in graphite (0.334 nm). In 1992, onions were obtained by *in situ* strong irradiation of carbon soot for the first time in a high-resolution transmission electron microscope [271]. Then in 1994, large-scale onion production (gram quantities) was realized for the first time by Vladimir Kuznetsov and coworkers, who used vacuum annealing of a nanodiamond precursor [139, 140].

When fullerenes are modified by introducing atoms of other chemical elements into the molecule, in case of onions the properties are changed by using another fullerene as an interstitial element, for example, C_{60} for interstitial insertion into the giant fullerenes C_{240}, C_{540}, and C_{960}. Figure 2.22a presents examples of the onion molecules. The onion properties were previously studied for the chain of Ar transition to Ar@C_{60} transition to Ar@C_{240}@Ar@C_{540} transition to Ar@C_{60}@C_{240} transition to Ar@C_{60}@ C_{240}@C_{540}. It is shown that, compared to the simple structure of Ar@C_{60}, the properties of the onion change quite significantly and are largely determined by the presence of several fullerene lattices. Such layered structures can be of great importance for the development of nanostructured materials science.

Figure 2.22 shows the TEM image of onions: (b) a large spherical onion [243], (c) a spherical onion with a hollow core [79], and (d) an onion with a metal particle inside [18]. As can be seen, onions can be characterized by different sizes (from 2 to 200 nm) and different number of layers; can have dense, hollow, or filled core; and can have different shapes (spherical or polyhedral). The main methods of their production are annealing of ultradispersed nanodiamonds, pyrolysis, arch-arc discharge between two electrodes immersed in water, chemical deposition, and electron beam irradiation. The fabrication method determines the morphology

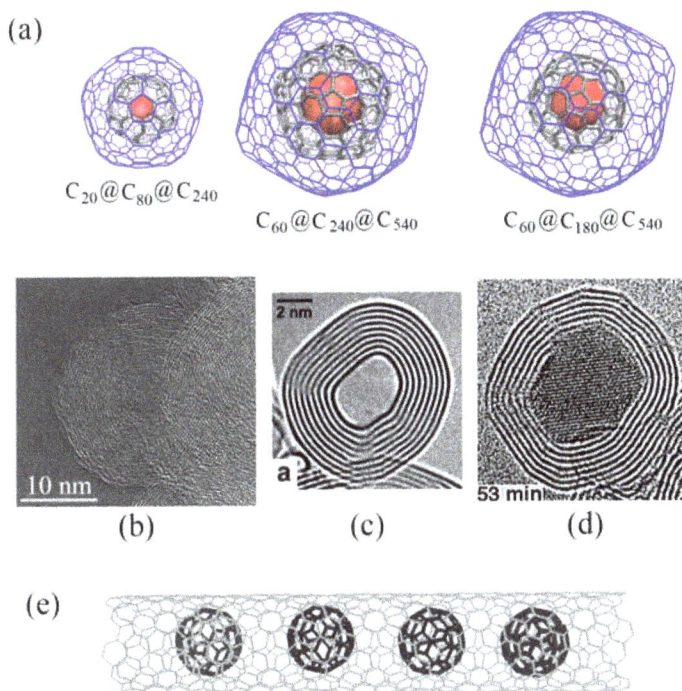

Fig. 2.22. (a) Examples of onion molecules. (b) Large spherical onion. Reprinted with permission from [243]. (c) Spherical onion with a hollow core. Reprinted with permission from [79]. (d) Onion with a metal particle inside. (e) Peapod: CNT with fullerenes inside. (b–d) are reprinted from [18].

and properties of the obtained onion. The unique 0D multi-layered sp^2-hybridized carbon shell structure of onions results in high mechanical resistance and conductivity. Their surface can be chemically functionalized to maintain their properties.

The structural diversity of carbon has been confirmed by the discovery of nanopeapods, a hybrid structure consisting of fullerene molecules encapsulated in single-walled CNTs [98, 132, 164, 257]. CNTs with appropriate diameters host fullerene molecules distributed along the nanotube axis. Not only fullerene molecules were found inside the CNTs, but also diamond-like molecules, fullerenes modified by other elements. An example is presented in Fig. 2.22e. Interestingly, the encapsulated molecules can rotate and move inside the CNT, which allow us to use it for nanomechanical devices [6, 258]. It was found that C_{60} molecules inside single-walled CNTs with a wide range of diameters (1.25–2.71 nm) can form ten different packing arrangements

[100, 101]: a chiral array; a double helical array [123]; a helix array [252], to name a few [41, 42]. In 1998, peapods containing C_{60} cages were prepared by Smith et al. for the first time using pulsed laser vaporization of graphite [257]. CNTs obtained using catalytically assisted arc evaporation are filled with fullerenes from C_{36} to C_{120} with the size defined by the nanotube diameter: narrower tubes contain smaller fullerenes [256].

Several studies have addressed the properties of CNT peapods for getting an understanding of their electronic, physical, and optical properties [103, 190, 228]. Commonly, the pepapod properties depends on the fullerene type, density, relative orientation to the nanotube and mutual arrangement of the fullerenes.

Strain can considerably affect the band gap and conductive properties of the peapods [93].

The diameters of the CNTs for peapods with C_{60} fullerenes should be in the range from 1.3 to 1.5 nm. In peapods of the optimal diameter, neither the CNT nor the fullerenes experience structural distortions that increase the energy and destabilize the structure. If the CNT diameter is reduced, then when fullerenes are inside, both the CNT and the fullerenes are deformed and the system becomes unstable. For CNTs with an optimal diameter of 1.3–1.5 nm, the potential curve of interaction with C_{60} has a single symmetrical minimum, which falls exactly on the tube axis, where the fullerenes are located. And for a tube with $D = 2.04$ nm, this curve has minima near the tube walls and a maximum on the axis. Consequently, in tubes of large diameters ($D > 1.5$ nm), the particles will shift from the tube axis and stick to its walls, which will disrupt the necessary linear periodicity in the distribution of fullerenes.

Initially, most of the C_{60} molecules form linear arrays with a distance between fullerenes about 0.3 nm, equal to the interplanar distance in graphite. Some pairs of fullerene molecules can apparently connect. It is assumed that at higher temperatures, fullerenes inside the CNT can combine into a stable corrugated nanotube inside the original CNT peapod. Remarkably, using a high-resolution tunneling microscope, the translational motion of fullerenes along the CNT and their rotation were observed in real time [257].

Peapods are conductors, where the carriers are distributed both along the tube and along the chain of fullerenes. It is important that for the symbiotic structure (in contrast to pure tubes) there is a noticeable change in the electron density along the peapod axis, which is determined by the periodicity in the arrangement of C_{60} fullerenes.

2.1.7 Graphyne and Graphdiyne

The appearance of 2D structures results in the development of a new field in materials science. The special properties of graphene and graphane raised an interest in the synthesis and investigation of other 2D carbon materials. Recently, carbon allotropes of various combinations of sp-, sp^2-, and sp^3-hybridized carbon atoms, for example graphynes, have also attracted considerable attention.

Graphyne is a monolayer of sp^2- and sp^3-hybridized carbon atoms arranged in a crystal lattice. Depending on the number of acetylenic bonds between carbon atoms, graphyne is arranged into graphyne (one acetylenic bond), graphdiyne (two acetylenic bonds), graphtriyne (three acetylenic bonds), etc. [53]. In the first theoretical work devoted to the study of structural types of graphyne, three indices were used for the designation of graphyne type in reference to the number of carbon atoms in the smallest carbon ring [20]. Further, the same authors introduced the designations by Greek letters, chosen concerning the set of symmetry operations for hexagonal graphene: α-graphyne is closest, etc. [51]. The same notations were used further, but additional indices were introduced to describe the same structural symmetry of graphynes [22]. To date, there is an uncertainty in the structural names of graphyne. The best-known form of graphyne is γ_1-graphdiyne [163]. Figure 2.23 presents the structure of well-known graphynes and graphdiyne.

As well as for many other carbon polymorphs, graphyne was first proposed in theoretical works [57, 58]: it was proposed to synthesize graphyne by the polymerization of tetraethynylethene or phenylacetylene having carbon cage fragments close to the structural type of graphyne. In a series of experiments, molecular fragments of graphyne of almost all structural types have been synthesized [55, 111, 173, 196, 239]; however, only the large-sized graphdiyne and graphdiyne nanotubes has been successfully synthesized to date [148, 149].

Historically, among the very rich variety of possible graphene-like $(sp + sp^2)$ networks, the family of so-called graphdiyne has been considered separately. The first representative of graphdiyne was proposed in 1997 [94]. Typically, these structures can be described as graphene networks, where all acetylene bonds are replaced by diacetylene bonds $-C{\equiv}C-C{\equiv}C-$. Thus, the term graphdiyne comes from the term graphyne. In addition, one can assume the existence of such exotic hybrids as graphene/graphdiyne, graphyne/graphdiyne, or graphene/graphyne/graphdiyne, etc.

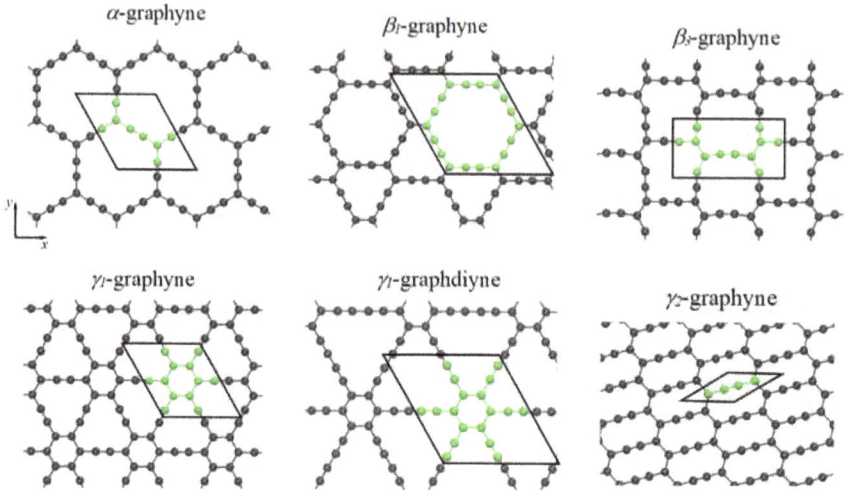

Fig. 2.23. Atomic structure of the part of the simulation cell for five configurations of graphyne (α, β_1, β_3, γ_1, γ_2) and γ_1-graphdiyne. The β_3-graphyne and γ_2-graphyne have orthorhombic anisotropy. All other structures have hexagonal anisotropy. The atoms in the periodic cells of graphyne and graphdiyne are colored green.

The mechanical, electronic, optical, and thermal properties of graphynes have not yet been experimentally determined due to the limited number and size of samples. High strength, attractive electronic properties, light weight, etc., were shown only by simulation. At room temperature, graphyne possesses high carrier mobility [116]. Like graphene, graphyne also has Dirac cones with opposite directions in its energy band, but the band gap increases with the number of acetylenic bonds in the periodic cell [106, 192]. The optical properties of halogenated graphyne shows that halogenation could effectively modulate the band gap and thus affect the optical properties [105]. The presence of the acetylenic bonds in graphyne leads to a significant reduction in fracture stress and Young's modulus, and the degree of decrease is proportional to the percentage of acetylene bonds [304].

Graphyne has been considered as a candidate for application as ultraviolet light protectors [24], transistors [244, 295], catalysts [283], anode material for future-generation Li-ion batteries [64], etc. The band gap controllability of graphyne offers great application potential [148, 277, 295]. Much like graphene, graphyne can be used to desalinize water: 100% ion uptake in seawater using pure graphyne was predicted by simulation [292].

The results of theoretical calculations show that graphdiyne due to its unique structure with mixed hybridization is a promising filter for dangerous gases, and can also be used in lithium-ion batteries, and for hydrogen storage. As for experimental production, in 2010 graphdiyne was successfully synthesized on the surface of copper foil [148, 302] (see Fig. 2.24). The resulting film, about 1 μm thick, has excellent semiconductor properties. However, obtaining graphdiyne with a controlled structure and specified properties is still of great importance and represents a serious problem.

Water molecules or ions can penetrate through graphynes at an operating pressure of up to 250 MPa, since the structure has extremely small nanopores. Moreover, graphynes demonstrate high energy barriers for water molecules and salt ions. Some types of graphynes can stop the movement of all ions in seawater, including Na^+, Cl^-, Mg^{2+}, K^+ and Ca^{2+}, and provide water permeability two orders of magnitude higher than other materials.

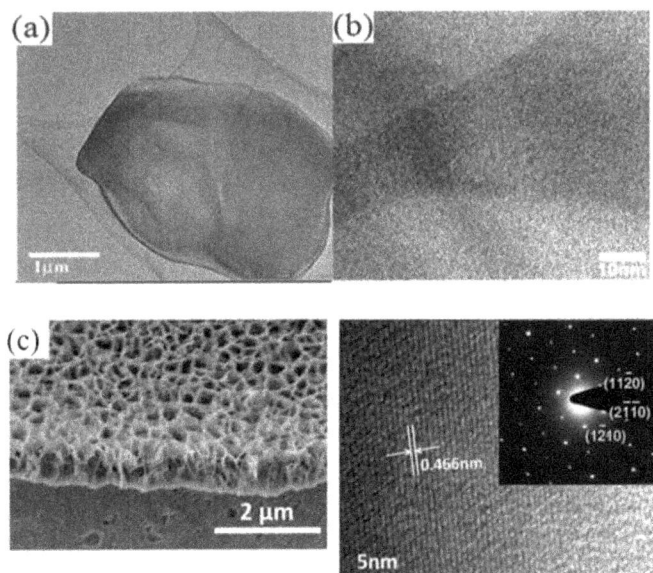

Fig. 2.24. The TEM images of graphdiyne films grown on the surface of copper foil: (a) low-magnification TEM image showing graphdiyne film, (b) high-magnification TEM image showing the edges of film regions of (a). Reprinted with permission from [148]. (c) SEM image, HRTEM image and corresponding SAED pattern of graphdiyne nanowalls. Reprinted with permission from [302].

2.2 Graphene-Based Materials

Currently, the technology of fabrication of new bulk carbon nanomaterials consisting of different types of carbon polymorphs, such as fullerene molecules, CNTs, crumpled graphene flakes, etc., is actively developing. The possibility of the existence of high-strength 3D carbon forms on the basis of covalent bonds only, such as schwarzschites, polybenzenes, and porous graphenes, has been also theoretically demonstrated. These materials can be classified based on the type of chemical bonds, as well as the number of nearest neighbors with which each atom forms a covalent bond.

The search for such new 3D graphene-based structures and the analysis of their properties such as conductivity, field emission, strength, and chemical activity, to name a few, is of great importance. This opens up new possibilities of their application as supercapacitors or electrodes in energy conversion devices, catalysis, separation and accumulation of gaseous substances, as protective coating, etc. To date, cheap and and environmentally friendly methods for the synthesis of such structures are being developed, and it is necessary to carefully study the structural stability of such materials, their properties and the effect of the external factors on their properties. In addition, composite materials based on different carbon polymorphs can be developed.

2.2.1 *Fullerite*

Under certain conditions, fullerene molecules, such as C_{60}, are ordered in space, located at the nodes of the crystal lattice. This system is a typical molecular crystal, in which the interaction between carbon atoms within a C_{60} molecule is significantly stronger than between atoms of neighboring molecules. Individual C_{60} molecules should be considered as inert molecules that retain their individuality in interactions with other similar molecules.

The crystal of fullerite C_{60} has a cubic structure with a face-centered cubic (FCC) lattice, a lattice constant of 14.2 Å and a density of 1.65 ± 0.03 g/cm^3 (see Fig. 2.25a, b). At room temperature, hexagonal (HCP) packing of molecules is also observed, although the FCC packing is preferable. The substance is stable in air, does not melt or decompose up to 360°C, and above this temperature it begins to sublimate. At room temperature, the centers of the molecules form a regular FCC crystal lattice, but the molecules themselves freely rotate around their centers. When the temperature decreases to 250–260 K, a first-order phase transition occurs: the free rotation of the molecules ceases, and they are oriented relative

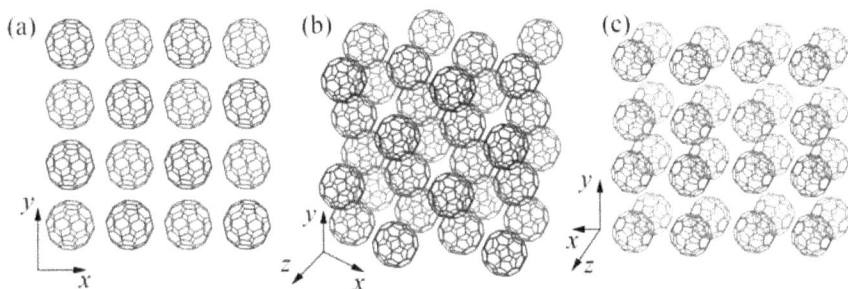

Fig. 2.25. (a, b) Fullerite with FCC packing in a projection to xy and in a perspective view. (c) Fullerite SC (simple cubic) packing.

to each other where their centers are slightly displaced from the positions corresponding to the ideal cubic arrangement, and a change in the crystal structure of fullerite occurs. The low-temperature phase (<260 K) has a primitive cubic lattice (see Fig. 2.25c).

Fullerites are semiconductors with a band gap of 1.5 to 1.95 eV. Since fullerites are fairly loose structures, the change in specific electrical resistance under pressure was studied first. Fullerites have photoconductivity under optical irradiation. Fullerenes in crystals are characterized by relatively low binding energies; therefore, in such crystals, phase transitions are observed even at room temperature, leading to orientation disordering and unfreezing of the rotation of fullerene molecules. C_{60} crystals doped with alkali metal atoms enter the superconducting state in the range from 19 to 55 K (a record temperature range for molecular superconductors).

A very important property of the fulleranes is that they can be transformed to fullerides if electron-donationg alkali atoms are introduced into the lattice. As a result, the whole structure transforms to a metal state, where donated electrons move between fullerenes. Fullerides demonstrate superconductivity. In [117] it was found for the first time that alkali-doped fullerides show a transition from a Mott insulator to a superconductor. Figure 2.26 demonstrates the crystal structure and electronic phase diagram of cubic fullerides.

2.2.2   Carbon Nanotube Bundles

The strong van der Waals bonding between nanotubes during their fabrication results in the formation of CNT conglomerates in solution. Very often, single-walled CNTs form densely packed bundles. As for fullerites, there is a special packing of the CNTs in the bundle. Figure 2.27 presents

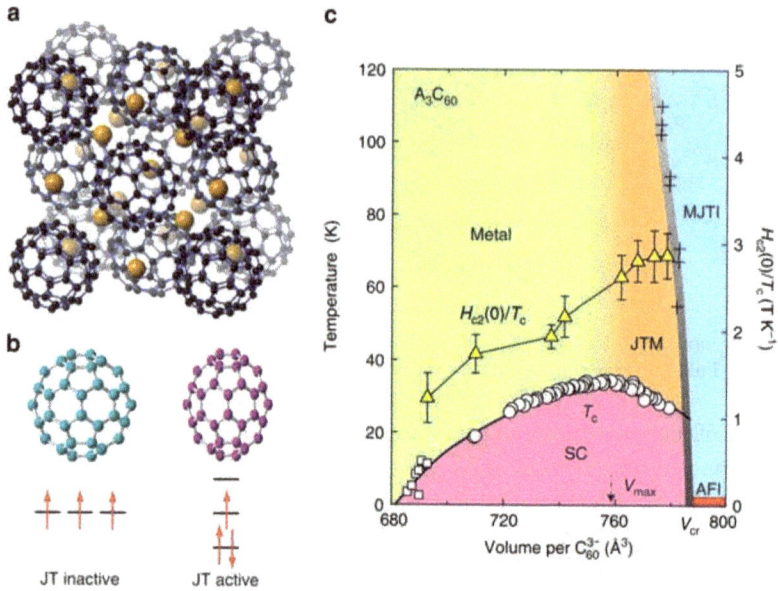

Fig. 2.26. (a) Crystal structure of FCC A_3C_{60} (A: alkali metal). Orange and black spheres represent A and C atoms, respectively. The C_{60}^{-3} anions adopt two orientations related by 90° rotation about the [100] axis. Only one is shown at each site. (b) Schematic structures of C_{60}^{-3} anions and molecular t_{1u} orbitals. (c) Electronic phase diagram of cubic fullerides. Squares and circles are the superconducting (SC) transition temperature T_c for FCC anion packing. Reprinted with permission from [117].

examples of the organisation of CNTs in bundle in a projection to the xy plane and in perspective view: simple and hexagonal packing.

It was shown that for hexagonally packed arrays of tubes, there are two stable states with different potential energies: an opened cross-section of CNTs and collapsed CNTs (which depends on the CNT diameter). Nanotubes collapsed along their axis have been studied both experimentally and theoretically, and the process of their collapse has been called the "domino process".

The structural characteristics of CNTs determine the possibility of transition from one stable state to another. The chirality of CNTs has almost no effect on the stability of the open configuration. The most important parameter in this case is the CNT diameter: for diameter of less than 1 nm, nanotubes remain practically of an ideal cylindrical shape, while nanotubes with a diameter greater than 2.5 nm collapse due to the action of van der Waals forces. The energy of the collapsed state of the

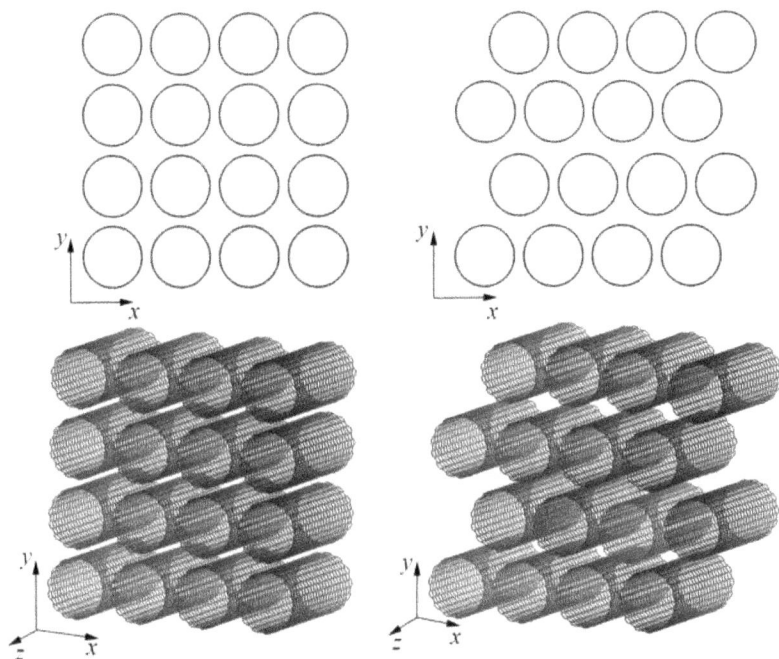

Fig. 2.27. Examples of the organization of CNTs in bundle in a projection to xy plane (above) and in perspective view (below): simple (left) and hexagonal packing (right).

nanotube is quite high and decreases nonlinearly with increasing diameter (Fig. 2.28). The collapse of CNTs occurs due to the action of van der Waals forces, while in the open state the nanotube is stable due to elastic forces. The ratio of the two forces determines whether the open or collapsed state will be stable. For nanotubes of small diameter, the energy of elastic interaction (open state) predominates, and for nanotubes of large diameter, van der Waals forces (collapsed state) predominate.

Due to their unique hollow tubular structure, large SSA, and good chemical and thermal stability, CNTs are considered as a promising candidate for gas adsorption. However, experimental results on hydrogen storage in carbon nanomaterials differ by several orders of magnitude. In 1997, it was reported that single-walled CNTs could store 10 wt.% hydrogen at room temperature, and the feasibility of meeting the benchmarks set for onboard hydrogen storage systems was predicted. However, after a few years, data began to appear on very low hydrogen storage capacity in CNTs, in particular those experimentally obtained at room temperature.

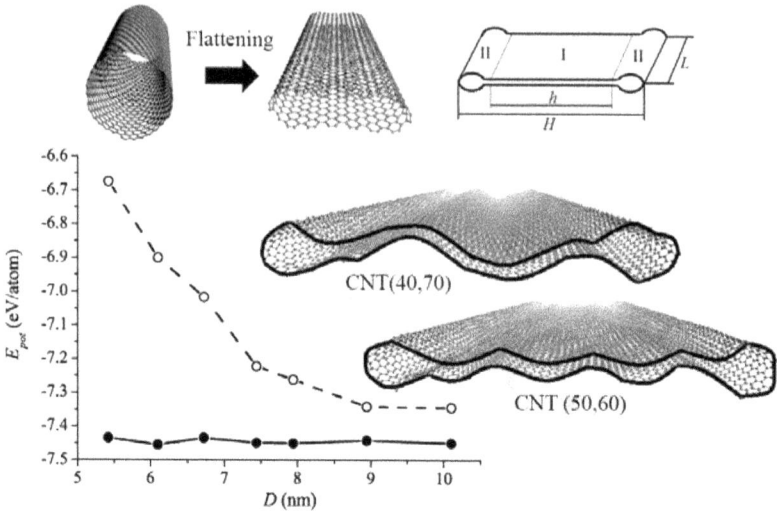

Fig. 2.28.   The potential energy per atom as a function of CNT diameter D for region I (solid line) and region II (dashed line). Insets: collapsed CNT(40,70), CNT(50,60).

It has been shown that the application of CNTs for hydrogen storage is a rather controversial area of research, since the mechanisms of hydrogen storage in CNTs remain unclear. Thus, the amount of hydrogen that can be stored in CNTs is less than 1.7 wt.% at a pressure of about 12 MPa and at room temperature.

2.2.3   *Crumpled Graphene Aerogel*

Novel carbon nanostructures composed of crumpled graphene flakes represent a new class of structures that have attracted much attention due to their unique mechanical and physical properties. Such structures were first observed in experiments from fullerene soot, a light porous carbon that forms on the walls of an evaporation chamber during the synthesis of fullerenes. It was believed that the material consists of fragments of fullerene-like carbon in which pentagons and heptagons are randomly distributed over a hexagonal network, forming a continuous curvature [96]. An example of such a structure is shown in Fig. 2.29a.

The study of crumpled graphene, 3D graphenes or graphene aerogels and some other 3D graphene-based materials is the latest trend in modern nanoscience. One of the latest findings is that crumpling eliminates the bonding of layers, which prevents the undesirable transition to graphite.

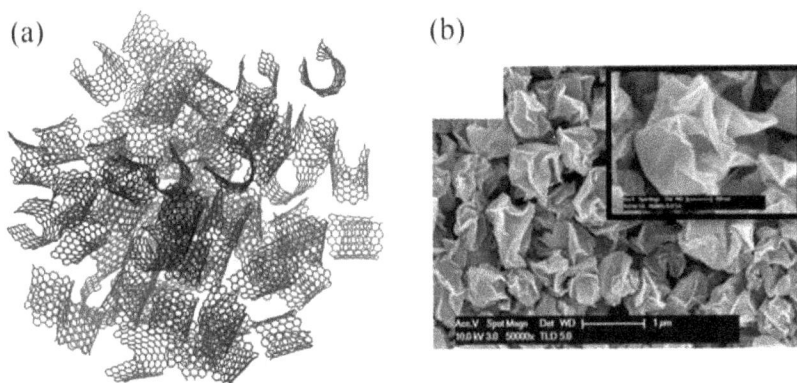

Fig. 2.29. (a) Schematic of crumpled graphene. (b) Crumpled graphene balls obtained by SEM. Reprinted with permission from [114].

Morover, formation of this structure can be carried out in a controlled or uncontrolled way which will affect their properties.

Figure 2.29b shows an example of crumpled graphene obtained using SEM [114]. Obviously, such a structure will be characterized by a large number of micro-, meso- and macropores, a large SSA and new properties due to the presence of a large number of folds.

Graphene aerogel (GA) is a highly porous 3D carbon material with skeleton walls that can be considered as ultrathin graphene films. High porosity, ultralight weight, and relatively simple fabrication tecniques result in a great demand for GAs in different applications ranging from pollutant adsorption to energy storage, catalyst support, supercapacitor, thermal insulation, etc. For the first time, carbon aerogel was synthesized through the pyrolysis and carbonization of organic gel in 1990 [203]. Since then, many different fabrication methods were developed for GA: for example, the self-assembly or gelation of graphene oxide suspension via hydrothermal reduction, chemical reduction or direct crosslinking of graphene, 3D printing, chemical reduction, and chemical cross-linking, to name a few.

Figure 2.30 presents GAs of different possible morphology: honeycomb, lamellar, cellular, and GA with randomly oriented flakes (can be also considered as crumpled graphene). Here, both model representations of experimentally synthesized GAs are presented [156, 221, 226, 302].

The GA morphology will considerably affect their properties, for example, some structures have ordered pores, while pores for GAs with

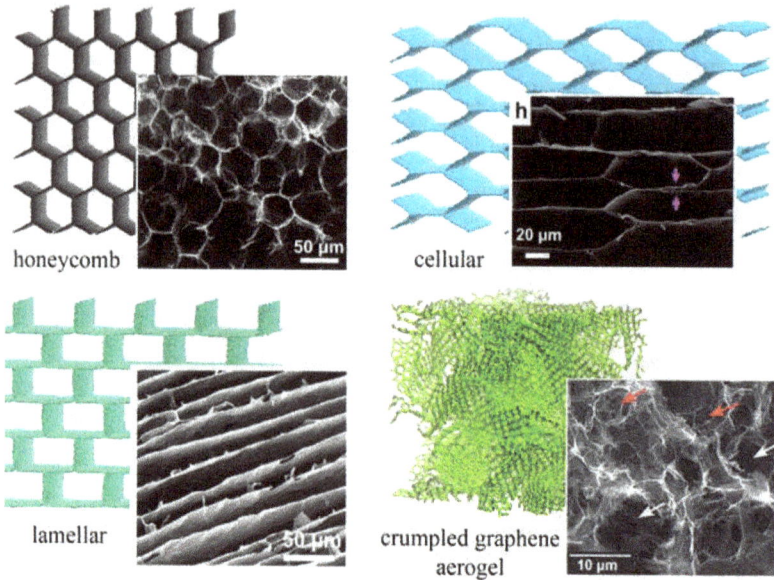

Fig. 2.30. Model representation and examples of the experimentally synthesized GAs: honeycomb GA, lamellar GA, cellular GA, and GA with randomly oriented flakes or crumpled graphene aerogel. Reprinted with permission from [156, 221, 226, 302].

randomly oriented flakes cannot be controlled. The quality of flakes and the density of the compaction of the skeleton are crucial for heat conduction [156, 284]. The thermal conductivity varies in a wide range from very low 4.7×10^{-3}–5.9×10^{-3} W/m·K due to the larger thermal contact resistance at the interfaces between adjacent graphene nanoribbons [288] to 28×10^{-3}–39×10^{-3} W/m·K [45]. It is still a great challenge to improve the thermal conductivity of such 3D graphene network to achieve a high through-plane thermal conductivity of more than 10 W/m·K. The structure optimization (pore structure, lattice structure, and microstructure) or functionalization (by chemical elements) of GAs is a promising way to improve the applicability of GAs.

2.2.4 *Diamond-Like Phases*

Carbon diamond-like phases (DLPs) are structures composed of carbon atoms, very similar to diamond by their morphology, with sp^3 or sp^2-sp^3 hybridization. In accordance with the hybridization, DLPs can be divided into two groups: (i) all atoms have the same sp^3 hybridization and are in crystallographically equivalent positions; (ii) the hybridization states of the atoms are also close to sp^3, but are not crystallographically equivalent.

An example is nanodiamonds, in which the carbon atoms have coordination numbers $k = 4$ characteristic of crystalline diamond and electron configurations close to sp^3. Nanodiamonds are divided into three families by origin: mineral (diamond-like hydrocarbon clusters isolated from oil), cosmic (meteorite), and artificial. Nanodiamonds are carbon nanomaterials that are quite heterogeneous in atomic structure and physicochemical properties, and the morphology of these structures can be quite different. For example, nanodiamonds can be observed as nanoneedles, nanofibers, nanowhiskers, etc. The transition from sp^3 to sp^2 hybridization can be accomplished in various ways, for example, as a result of heat treatment or compression. These materials are attracting attention as anti-friction materials and additives to oils, as well as for the production of metal-diamond hardening coatings. There are developments for the use of nanodiamonds in medicine and biology.

To date, various DLPs have been experimentally synthesized and theoretically studied, for example, cubic diamond, hexagonal diamond polytype (lonsdaleite), polymerized cubic fullerite C_{24}, and high-density carbon C_8 phase (examples are shown in Figure 2.31). One example is supercubane, the structure of which is obtained by crosslinking the carbon frameworks of cubane molecules linked together by carbon-carbon bonds in the directions of the cube diagonals. A nanodiamond phase of polymerized nanotubes (4,0) was proposed, and theoretical studies have shown that the structure of such a phase is covalently bonded nanotubes (4,0).

Figure 2.32a shows a diamond and three typical structures of diamond-like phases: fullerane, tubulane, and graphene-based DLP. Fulleranes are

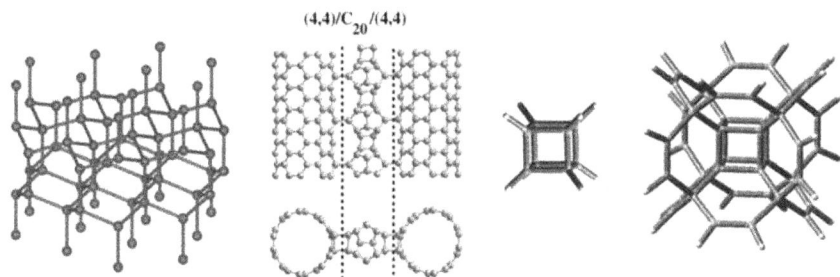

$(4,4)/C_{20}/(4,4)$

Fig. 2.31. (From left) Hexagonal diamond polytype (lonsdaleite), polymerized cubic fullerite C_{24}, high-density carbon C_8 phase (cubane and supercubane).

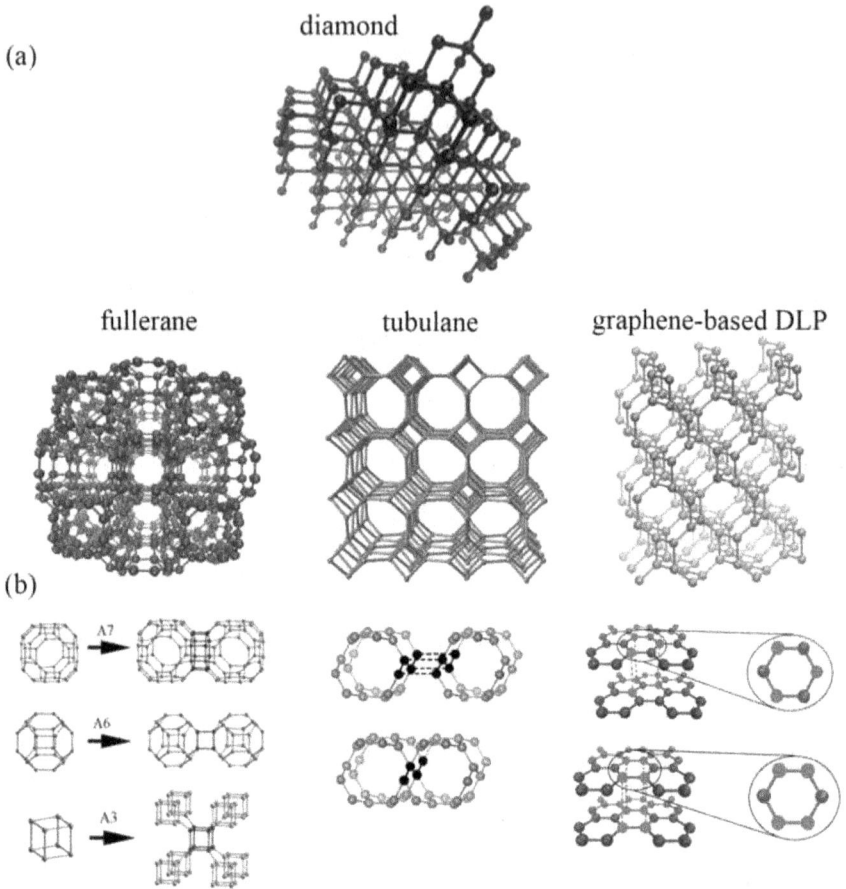

Fig. 2.32. (a) Diamond and three typical structures of DLPs: fullerane, tubulan, and graphene-based DLP. (b) The construction of DLPs from fullerene-like molecules, CNTs and graphene layers.

DLPs based on different fullerene-like molecules, tubulanes are DLPs based on CNTs, and the third class of DLPs is graphene-based DLPs. As can be seen, all the carbon-based structures are different, but each of them can be characterized by the same hybridization. Figure 2.32b shows examples of the construction of DLPs from fullerene-like molecules, CNTs and graphene layers.

Among the 35 currently known DLPs, each has its own geometry, structural parameters, and unique properties. As shown by *ab initio* calculations, such structures can be obtained by combining different carbon polymorphs. Several possible scenarios for the synthesis of DLPs can be described. As is known, strong compression at high temperatures can lead to the transformation of sp^2 carbon into sp^3 carbon. By compressing various carbon nanopolymorphs, various carbon phases are formed, including diamond, transparent superhard post-graphite phases, and superhard amorphous carbon with all sp^3 hybridization. All the details on the structure of the described DLPs are presented in [151, 232, 233].

Fulleranes can be composed of fullerene-like molecules C_4, C_6, C_8, C_{16}, C_{24}, C_{48}, which are similar to fullerenes, but have a smaller number of atoms and different shapes. Fulleranes are named as C - carbon and A or B - the way of interconnection between fullerene-like molecules. Single elements have sp^2 hybrid state, and when combined into 3D structure, they pass into a mixed $sp^2 - sp^3$ hybridization. Fullerene-like molecules, such as C_{24}, C_{48}, etc., can be connected into one structure in several ways: along an edge, along a tetrahedron, along a hexagon, and along an octahedron. Depending on the type of connection, the properties of the phase will vary greatly. Among ten fulleranes, eight (named CA1, CA3, CA4, CA6, CA7, CA8, CA9, CB) have cubic anisotropy, CA2 phase has hexagonal anisotropy and CA5 phase is tetragonal. All the stable fulleranes are presented in Fig. 2.33.

Tubulanes consist of nanotubes differing in chirality, length, connection method, and are represented by six types of nanotubes (2,0), (2,2), (3,0), (3,3), (4,0), (6,0). The limitation for the size of the CNTs is because combination of large CNTs results in the formation of van der Waals sp^2 hybrid state, and CNTs with small indexes become fully sp^3. Tubulanes are named as T - tubulane and A or B - the way of interconnection between CNTs. The same type of CNTs can form several different compounds, for example, a CNT(3,3) is the unit for three phases (TA2, TA8, TB), a CNT(4,0) for two (TA5, TA6). Among nine phases, four configurations (named TA1, TA3, TA5, TA6) are tetragonal, four (TA2, TA4, TA7, TB) are hexagonal, and one (TA8) is triclinic (Fig. 2.34).

Graphene-based DLPs can be obtained by crosslinking of graphene layers of different configurations. As in the case of fulleranes and tubulanes, graphene can be used to obtain different configurations of DLPs. Among eight graphene-based DLPs, LA1 and LA4 are cubic, LA2 is hexagonal,

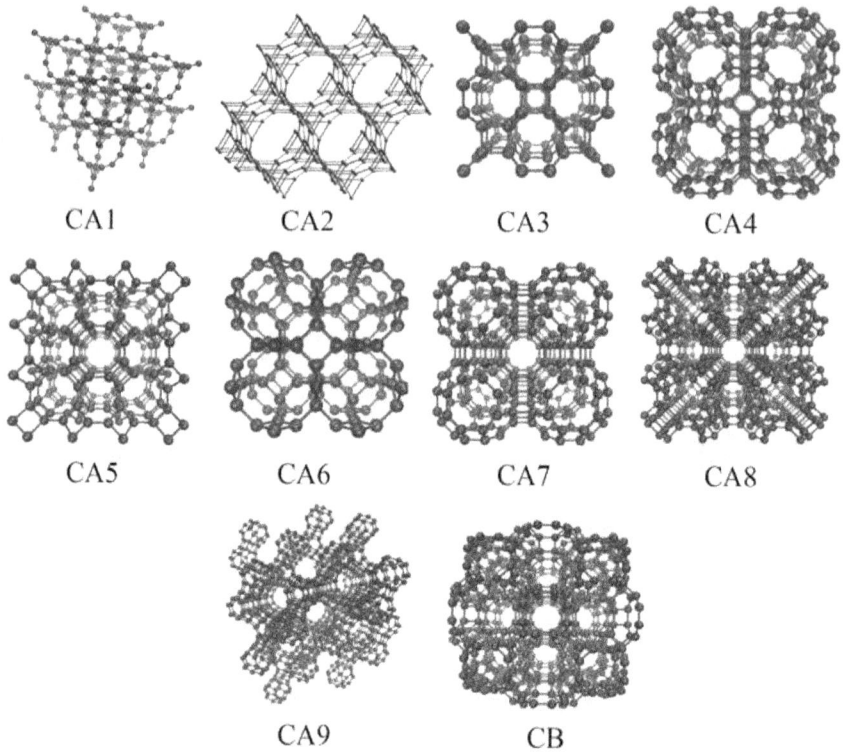

CA1   CA2   CA3   CA4

CA5   CA6   CA7   CA8

CA9   CB

Fig. 2.33. Fulleranes.

LA3 and LA8 are tetragonal, and LA5, LA6, LA7 are rhombic. In addition, they can also be divided by the type of connection between graphene layers: three bonds (LA1, LA2, LA4, LA45, LA6) and four bonds (LA3, LA7, LA8) (see Fig. 2.35).

The most preferred method for synthesizing such DLPs is strong uniaxial compression of various types of graphite along axes perpendicular to the graphene layers. One of the graphene-based carbon nanostructures, lonsdaleite (other names LA2, 3D (3.0), hexagonal diamond, 2Hdiamond), was first obtained in 1967 at static pressures above 13 GPa and temperatures above 1273 K, and was also found in the environment as solid particles in a meteorite. Currently, samples containing both lonsdaleite and diamond can be synthesized in laboratories under high pressures and temperatures.

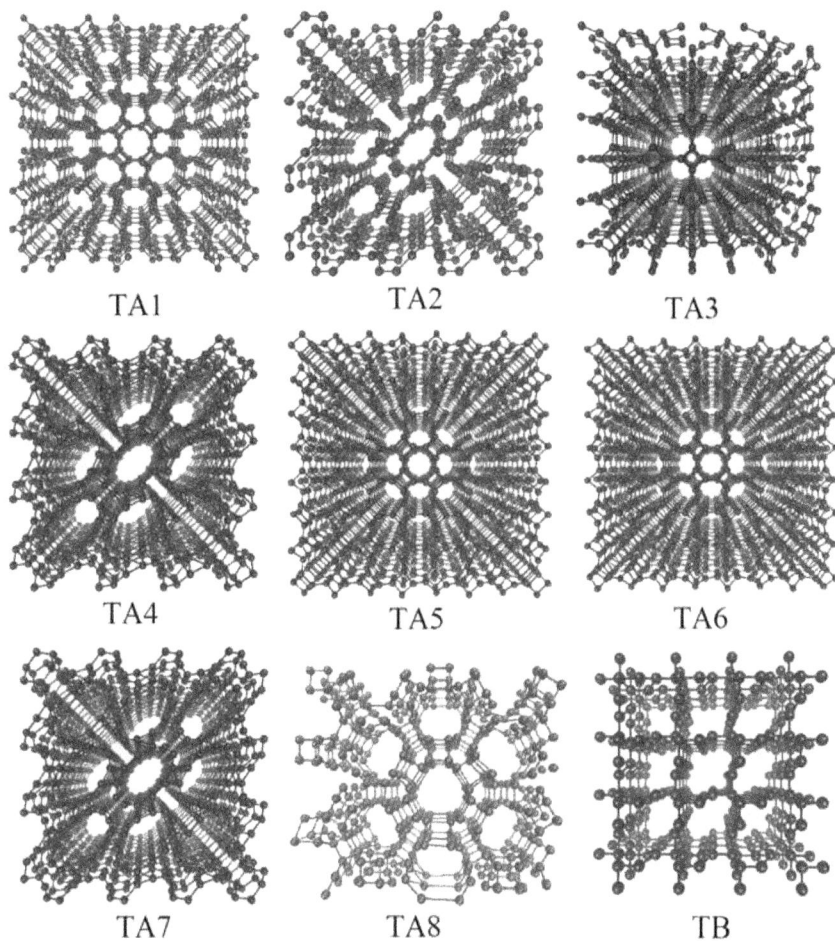

Fig. 2.34. Tubulanes.

A number of structures have been obtained by chemical synthesis, for example, the C_{120} dimer, consisting of C_{60} fullerenes, and the C_{130} dimer, consisting of C_{60} and C_{70} fullerenes.

The most important applications for DLPs are surface coating in biomedicine, protection of the surface from external influences, and anti-friction coatings.

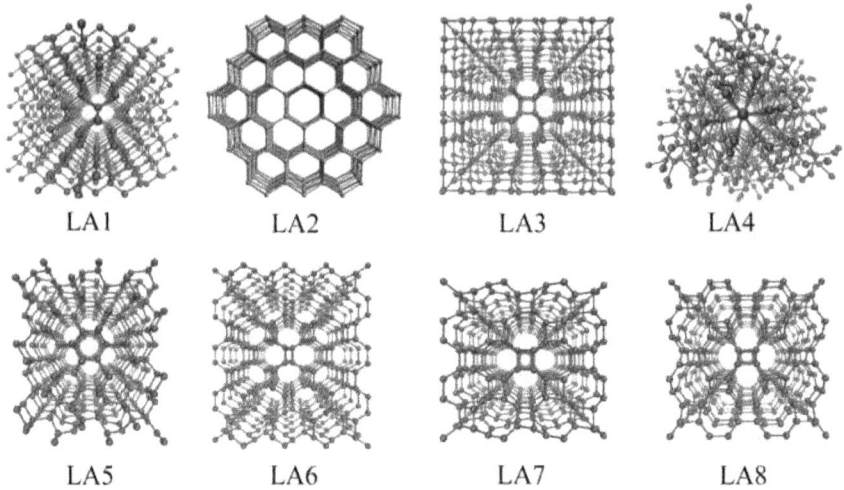

Fig. 2.35.   Graphene-based DLPs.

2.3   Graphene-Based Composites

Graphene as a reinforcement for metals has attracted a lot of attention in the last few years due to its high Young's modulus, fracture strength, large surface area, low density, and excellent thermal and electrical conductivity. Graphene-based composites with improved mechanical properties are highly important for the aerospace, automotive, shipbuilding and oil industries, as well as in many other areas. In recent decades, much attention has been paid to the development of composite materials based on carbon polymorphs such as graphene and CNTs, since they can be effective reinforcing elements due to high strength and flexibility. It is well known that CNTs in a polymer matrix improves mechanical, electrical and thermal properties of such composite. Graphene-based composites can demonstrate the wide variety of mechanical and physical properties for different morphologies and different composite components. The most widespread are the metal/graphene and polymer/graphene composites. Here, metal/graphene composites will be considered.

Due to their excellent properties Al, Cu, Ni, ferrite, and Ti are commonly considered as the widespread metal matrices. Aluminum is a lightweight metal with good corrosion resistance and formability, which can be used for numerous applications, but its main disadvantage is low strength. Copper is also lightweight, low in cost, with high electrical and thermal conductivity, but the corrosion resistance is lower than required. Nickel has great corrosion and wear resistance associated with superior resistance to thermal oxidation and high strength. The strengths of Fe and Ti are quite high by themselves; however, they can be further increased by graphene reinforcement.

It is well-known that different metals have different adhesion energy, solubility, and catalytic properties for interaction with carbon itself and carbon polymorphs [13]. For Cu, Ag, and Au, there is electronic decoupling between metal and graphene, which results in quite strong interaction. The distance between metal surface and graphene is close to 3 Å, which is characteristic of van der Waals interaction. Metals such as Co, Ni, Ru, Rh, Ti and Re show even stronger interaction with graphene and shorter distances. Thus, Al, Cu, Ag, Au, and Pt have weaker cohesion with graphene, while Co, Ni, and Pd have stronger cohesion. Note that interaction energy can depend on the size of graphene sample, presence of defects, curvature of graphene, and the interaction between single or coupled C atoms and the metal surface can differ from the interaction between the graphene layer and metal surface.

Nowadays, graphene/metal composites can be obtained by different methods, such as powder metallurgy, mechanical ball milling and hot rolling, or even via the growth of a graphene layer on the metal surfaces. However, homogeneous distribution of graphene inside the metal matrix is very important for composite properties, but cannot be easily achieved.

Let us first consider two different graphene types: (i) planar graphene layer or flake and (ii) crumpled/wrinkled graphene. As was previously shown, these structures will have very different mechanical properties. The most spread morphology is a metal matrix reinforced with graphene layer. The idea to reinforce the metal matrix by the graphene appeared right after the first successful experiments on graphene exfoliation. Mechanical properties of metal matrix increased considerably, due to the interaction of dislocations with graphene.

Figure 2.36 presents a schematic representation of the variation of the composite morphology. Metal atoms are not shown, only graphene flakes. The morphology of the graphene-reinforced composites can be very different depending on the (1) shape of graphene flakes, (2) graphene distribution, and (3) the amount of graphene in the metal matrix. The composite properties will considerably depend on the graphene distribution, shape, size, and other different parameters that define the whole morphology.

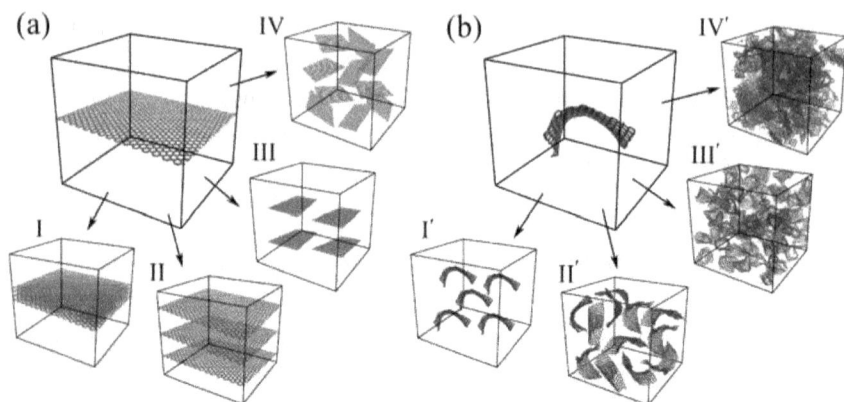

Fig. 2.36. Schematic of metal/graphene composites with (a) planar graphene and (b) crumpled graphene. Metal atoms are not shown. Reprinted with permission from [13].

The composite morphologies shown in Fig. 2.36 are schematic and the sizes and shapes of graphene reinforcements are shown randomly. Four examples are presented in Fig. 2.36a: I, II – several graphene layers with different interlayer distances, III – several small flakes, monotonously distributed in a metal matrix, and IV – several graphene flakes randomly distributed in the metal matrix. Here, for simplicity, small graphene layers will be called graphene flakes and large-area graphene will be called graphene.

The problem with graphene is that it is unstable in the planar form. The graphene tends to agglomerate and stack to multi-layered flakes or even to graphene networks. Figure 2.36b presents a similar schematic of the composite structure as for planar graphene: one graphene flake, several, and a graphene network. A graphene network will show different deformation behavior in comparison with the separate flakes.

However, the analysis of metal monocrystal reinforced with graphene is a bit far from real experiments. Commonly, the polycrystalline structure with different grain sizes should be considered. Graphene flakes can be distributed differently and can affect the resulting mechanical properties.

One of the recently developed type of graphene/metal composites is the mixture of graphene flakes and metal nanoparticles. Graphene/metal hybrid systems were already obtained for noble metal nanoparticles (Au, Ag, Pt, and Pd) [25, 241] and the transition metal nanoparticles (Ni, Co, Cu) [109]. Such hybrid graphene/metal systems can be synthesized by the direct growth of the nanoparticles on the graphene surface and solution mixing—the mixture of graphene flakes and pre-synthesized nanoparticles. In both approaches, graphene and metal nanoparticles can be bonded either by chemical bonding or van der Waals interactions. Despite various strategies for mixing nanoparticles and graphene precursors being developed to date, many unsolved issues remain: how to better assemble nanoparticles on graphene, control morphology and density of the composite, and improve properties according to practical requirements.

Figure 2.37a presents the results for the new approach to obtain the metal/graphene composite with graphene network filled with metal nanoparticles by hydrostatic compression. The initial structure of crumpled

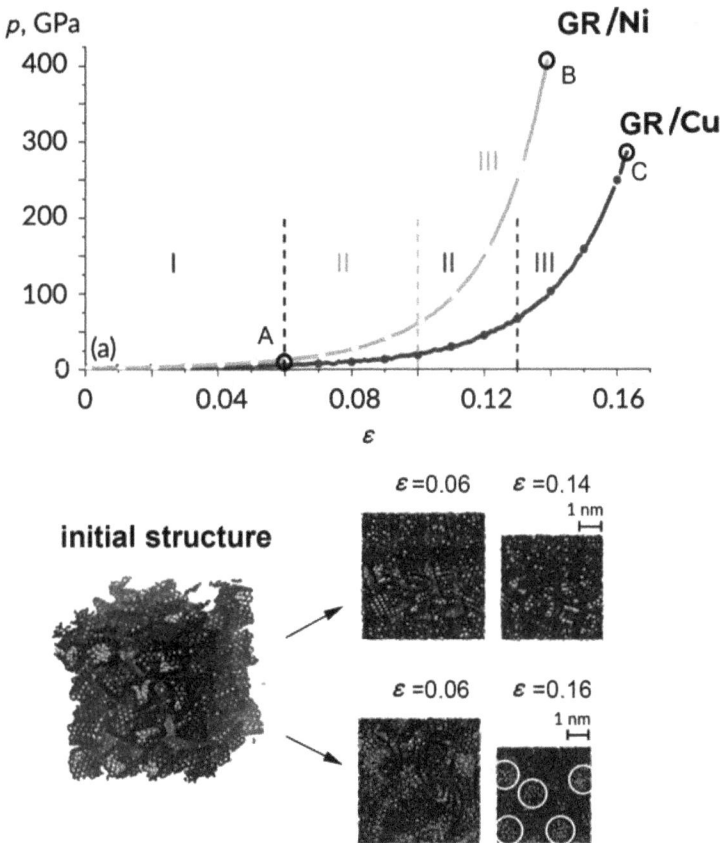

Fig. 2.37. (a) Pressure-strain curves for graphene/metal composite. (b) Initial structure of the composite composed of graphene network with the pores filled with metal nanoparticles. Transformation of the initial structure to composite under compression.

graphene with Ni nanoparticles inside the pores of crumpled graphene is shown. Graphene flakes are randomly rotated and filled with Ni or Cu nanoparticles. The fabrication technique is high-temperature hydrostatic compression. Moreover, the temperature of hydrostatic compression should be close to 0.7 T_{melt}^{NP}, where T_{melt}^{NP} is the melting temperature of the metal nanoparticle of the given size.

Figure 2.37b presents the pressure-strain curves during hydrostatic compression of two samples – graphene network filled with Ni nanoparticles (GR/Ni) and the same graphene network filled with Cu nanoparticles (GR/Cu). Snapshots of the composite structure during compression are also presented.

The size of nanoparticles is one of the key factors: if the surface area of metal nanoparticle is comparable with the SSA of graphene flake, the formation of a strong graphene network is prevented, which complicates the composite fabrication [138]. The second important factor is the metal of the nanoparticle, or the interaction energy between graphene and the chosen metal nanoparticle.

Chapter 3

Mechanical Properties

Graphene has a record-breaking Young's modulus with respect to in-plane tension, but its flexure rigidity is practically zero, which means that graphene can be easily bent. Wrinkling or rippling of graphene is a very well-known phenomena also found for other thin sheets and membranes. The structural state of graphene will considerably affect the mechanical and physical properties of graphene and other 2D carbon nanostructures.

The intrinsic mechanical properties of graphene can be found from the phonon frequencies under tensile and compressive strains. Raman spectroscopy is used to study phonon frequencies under uniform tension and hydrostatic pressure [174, 187]. It has been shown that tensile stresses lead to a decrease in phonon frequencies, while compressive stresses lead to an increase in the frequencies of phonon vibrational modes. The main experimental method for studying the mechanical properties of single-layer or multi-layer graphene is atomic force microscopy (AFM). AFM was used to study the elastic properties of a stack of graphene layers (less than five) on a silicon dioxide substrate [160]. The Young's modulus of the graphene stack was 0.5 TPa. A similar experiment was conducted with a graphene monolayer [145], and its Young's modulus was about 1 TPa. Another way to characterize the strength of a material is nanoindentation, which was used to study some properties of graphene [74, 216].

When studying mechanical properties, mechanical compression-tensile tests are usually carried out, bending deformations are analyzed, and elastic moduli are calculated – Poisson's ratio, Young's modulus, shear modulus, etc. Strength and elastic constants for all crystalline bodies considerably depend on the strength of interatomic forces. When considering tensile loads, the critical stresses that arise in a crystal are directly related to the strength of the chemical bond, and for covalent materials, as is known, the C-C bond is very strong. On the other hand, compressive strength is determined by completely different criteria – by the bending characteristics of the material. Elastic modulus correspond to a small strain when the deformation has an elastic, reversible nature.

Different simulation techniques are also widely used for the analysis of the mechanical properties, since it is difficult to implement such an experiment in practice. In this section, elastic properties, strength and deformation behaviour of graphene, diamane, graphynes and graphene-based composites will be considered.

3.1   Elastic Constants

3.1.1   *Graphene*

Graphene is a very strong, flexible, highly stretchable material with outstanding mechanical properties, such as an extremely high shear modulus of 280 GPa, a tensile Young's modulus of 1 TPa, and a strength of about 100 GPa. Undeformed graphene is an isotropic elastic medium, while uniformly deformed graphene is generally anisotropic. Graphene remains isotropic only under hydrostatic pressure (compression) $\varepsilon_{xx} = \varepsilon_{yy} \neq 0$, $\varepsilon_{xy} = 0$, and in the case of zero shear strain $\varepsilon_{xy} \neq 0$ and nonzero components $\varepsilon_{xx} \neq \varepsilon_{yy} \neq 0$, graphene is orthotropic.

Mechanical properties of any structures in the range of the elastic strain can be described by the well-known Hooke's law:

$$\sigma = E \cdot \varepsilon,$$

where E is the Young's modulus, which can be calculated from the analysis of stress (σ)-strain(ε) state.

Elastic constants can be defined from the tensile strain as

$$c_{11} = \frac{1}{WL} \frac{\delta^2 U}{\delta \varepsilon^2},$$

$$c_{11} + c_{12} = \frac{1}{WL} \frac{\delta^2 U}{\delta \varepsilon^2},$$

where L is the length of the sample and W is the width of the sample.

The Young's modulus can be obtained from the stress–strain curve. The slope of the linear section at the beginning of the curve represents the Young's modulus. Young's modulus of graphene is approximately independent of the temperature and chirality. Table 3.1 presents the values of the Young's modulus obtained from experiment and simulation. As can be seen there is a slight difference in the values of the elastic modulus,

Table 3.1. The values of Young's modulus E obtained from experiment and simulation [213].

Ref.	Exp./Sim.	E, N/m
[145]	exp.	340 ± 50
[69]	exp.	342 ± 8
[155]	exp.	349 ± 12
[297]	sim.	321; 285
[150]	sim.	320; 303
[224]	sim.	309; 303
[238]	sim.	248-330
[309]	sim.	270; 298
[35]	sim.	291; 324

which is the result of differences in the applied methods of investigation, studied graphene samples, temperatures, etc.

One of the interesting characteristics of graphene is that it demonstrates negative characteristics such as negative thermal expansion coefficient and negative Poisson's ratio at some special conditions. Materials with negative Poisson's ratio are very rare and called auxetics. Under tension auxetic materials become thicker in the direction normal to the applied force. This is due to the structure peculiarities, which allow such unusual behavior. Poisson's ratio indicates the degree of transverse compression of a material when loaded in one direction, and can be defined as

$$\nu = -\frac{\varepsilon_i}{\varepsilon_j},$$

where ε_i is the uniform deformation along some direction and ε_j is the corresponding deformation along the normal direction.

For example, in a certain range of deformations graphene becomes an auxetic, i.e., a material with a negative Poisson's ratio. Figure 3.1 presents the stability region divided in accordance with the sign of the Poisson's ratio. Here, the Poisson's ratio was calculated for different tensile strain inside the stability region. It can be seen that Poisson's ratio became negative for $\varepsilon_x = \varepsilon_y = 0.12$ for hydrostatic tension. Along the zigzag (x-axis) direction, the range of deformation which leads to auxetic behavior is much wider than along the armchair (y-axis) direction. Thus, there is a wide range of tensile strains which results in the auxetic behavior of graphene.

Auxetic behaviour was demonstrated in [91] for graphene crumpled by chemical functionalization (see the structures in Fig. 3.2a). Under tension of such crumpled graphene, the stress increases significantly as the applied

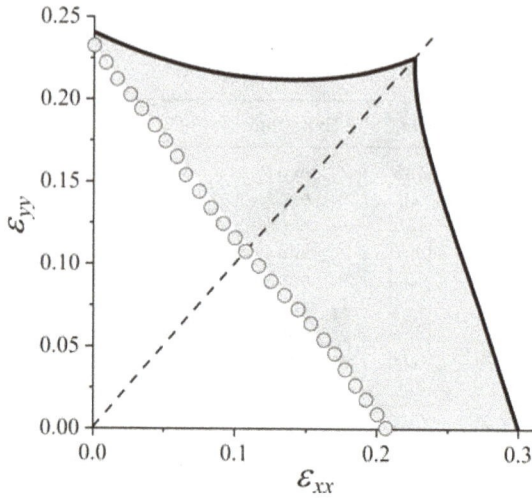

Fig. 3.1.   Stability region divided into regions with positive (white area) and negative (gray area) Poisson's ratio.

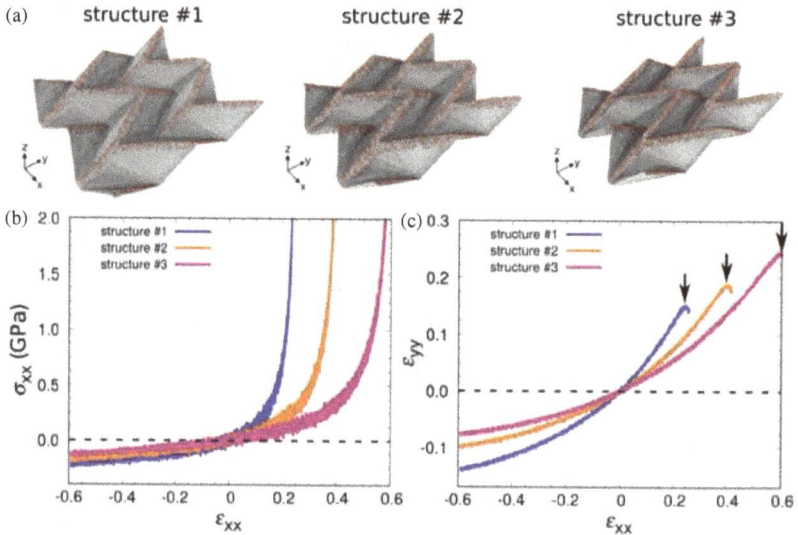

Fig. 3.2.   (a) Graphene Miura-ori structures (three different morphologies) with super-compliance, super-flexibility (both in tension and compression), and auxeticity. (b) Stress–strain curves. (c) Strain in the y-direction due to the applied strain in the x-direction. The arrows indicate the critical strain values at which the structures are flattened. Reprinted with permission from [91].

strain increases, and graphene is almost flattened. Crumpled graphene shows expansion/contraction in the y-direction when stretched/compressed in the x-direction (see Fig. 3.2c). The auxeticity is observed in an extremely wide range of applied strain. Note that the Young's modulus, flexibility, and Poisson's ratio of the Miura-ori structures depend on the geometry, which can be controlled by changing the folding angle.

3.1.2 *Diamane*

It was shown in experiment that diamane's stiffness is 1079±69 GPa [40]. Until now, the direct measurements of elastic constants of 2D structures had been very complicated. In [40], the layered structure composed of graphene grown on SiC and then compressed into diamane was considered. The stiffness of different layers was defined separately but still can affect the final value of the stiffness constant. The most suitable methods for the calculation of the elastic modulus of new structures such as diamane and graphynes are MD or DFT calculations. Tables 3.2 and 3.3 present the stiffness and compliance constants for diamane AA and AB with (sp^3 hybridization) and without (mixed sp^2-sp^3 hybridization) hydrogen, respectively.

The morphology of diamane slightly affects the values of stiffness constants. The stiffness constants for H-diamane AA and AB differ by 0.5–1%, and for diamane AA and AB without H the constants differ by 1.3%

Table 3.2. Stiffness constants (in GPa) for diamane [213].

Structure	c_{11}	c_{12}	c_{66}
Diamond	1098.16	128.48	750.0
Diamond [263]	1079.0	124.0	578.0
D-AA+H	808.5	47.6	381
D-AA+H [197]	1026	81	473
D-AB+H	804.5	47.6	378
D-AA	1225	65	589.8
D-AB	1209	82.6	563.3
H-diamondene [240]	1126.0	81.0	473.0

Table 3.3. Compliance constants (in TPa^{-1}) for diamane.

Structure	s_{11}	s_{12}	s_{66}
Diamond D-AA+H	1.25	−0.069
D-AB+H	1.25	−0.069	2.6
D-AA	0.8	−0.041	1.7
D-AB	0.8	−0.053	1.7

Table 3.4.   The values of Young's modulus E, Poisson's ratio ν, and shear modulus G of diamane [213].

Structure	E, GPa	G, GPa	ν
Diamond [198]	1144.6	534.3	0.07
D-AA+H	794	375	0.06
D-AB+H	795	375	0.06
D-AA	1187	564	0.05
D-AB	1182	550	0.07
H-diamane [177]	692	–	0.08

(c_{11}), 27% (c_{12}), and 4.7% (c_{44}). In [141], lonsdaleite films were considered by DFT and it has been found that elastic constants c_{11} and c_{12} depend on the interface orientation and the number of diamond layers.

For fluorinated diamane the stiffness constants obtained by DFT are $c_{11} = 499.4$ N/m (1107.8 GPa) and $c_{12} = 55.9$ N/m (124 GPa) [275]. These values are also very close to the stiffness constants obtained for H-diamane.

The Young's modulus of all chiral diamanes obtained from the stress-strain curves is 763 GPa. Thus for diamane, Young's modulus does not depend on the layer stacking, which is in agreement with [177, 213]. The orientation also has no effect on the elastic modulus.

Young's modulus, Poisson's ratio and shear modulus can be respectively calculated from the stiffness constants as:

$$E = \frac{c_{11}c_{22} - c_{12}^2}{c_{11}}, \quad \nu = \frac{c_{12}}{c_{11}}, \quad G = c_{66}.$$

It was found that Young's modulus are 805 GPa for D-AA+H and D-AB+H, 1220 GPa for D-AA, and 1203 GPa for D-AB [213]. Without hydrogen, stiffness constants and Young's modulus are larger, due to the formation of sp^3 bonds. Table 3.4 presents the elastic constants calculated in [213] with the methodology described in Chapter 4.

For diamane, layer stacking has no effect on Young's modulus. Poisson's ratio ν for all investigated diamanes is in the range from 0.06 to 0.07, which is close to that of diamond (or diamond-like films).

3.1.3   *Graphyne and Graphdiyne*

Tables 3.5 and 3.6 present the stiffness and compliance constants, respectively, obtained for graphene, graphane, five configurations of graphyne, and γ_1-graphdiyne calculated in the present work and from the literature (shown in brackets). Except for β_3- and γ_2-graphyne, the stiffness constants c_{11}

Table 3.5. Stiffness constants c_{ij} for five graphynes and γ_1-graphdiyne. The values in brackets are taken from the literature [4, 104, 152, 204, 205, 281].

Structure	c_{11}, N/m	c_{22}, N/m	c_{12}, N/m	c_{66}, N/m
		Hexagonal		
α-graphyne	87.44	-	68.53	8.70
	95 [104]	-	82 [104]	6.50 [104]
β_1-graphyne	126.90	-	71.63	27.43
	133 [104]	-	86 [104]	23.50 [104]
γ_1-graphyne	177.62	-	65.28	61.23
	198.70 [204]	-	85.30 [204]	60.0 [104]
γ_1-graphdiyne	138.40	-	61.97	38.22
	152.10 [4]	-	69 [4]	41.60 [4]
		Orthorhombic		
β_3-graphyne	151.60	100.13	10.53	18.173
γ_2-graphyne	369.98	118.48	76.43	84.98

Table 3.6. Compliance s_{ij} constants for graphene, graphane, five configurations of graphyne (α, β_1, β_3, γ_1, γ_2) and γ_1-graphdiyne.

		Hexagonal		
α-graphyne	29.60	-	-23.20	114.90
β_1-graphyne	11.57	-	-6.528	36.50
γ_1-graphyne	6.51	-	-2.39	16.30
γ_1-graphdiyne	9.04	-	-4.05	26.20
		Orthorhombic		
β_3-graphyne	6.64	10.06	-0.699	55.03
γ_2-graphyne	3.118	9.738	-2.012	11.77

and c_{22} are equal, therefore, in the table, instead of repeating the values in column c_{22}, there are dashes. The same is true for the compliance constants.

Among the considered hexagonal configurations of graphyne, the largest stiffness constant c_{11} is demonstrated by γ_1-graphyne, the lowest by α-graphyne. It can be seen from the tables that graphdiyne is a softer material than either graphyne or graphene.

Thus, it can be concluded that the inclusion of acetylenic linkages in the graphene lattice decreases the stiffness of the material, obtaining softer 2D materials.

Table 3.7 presents the comparison of Young's modulus, Poisson's ratio and shear modulus of graphene, graphane, the five graphynes, and γ_1-graphdiyne under study with literature data. The values obtained in this study for graphene and graphane are presented to support the reliability of the methodology employed. The experimental value of Young's modulus for graphene is 340 ± 50 N/m [145].

Table 3.7. The values of Young's modulus E, Poisson's ratio ν, and shear modulus G of hexagonal crystals (α-, β_1-, γ_1-graphyne, and γ_1-graphdiyne) obtained in the present work in comparison with literature [4, 5, 9, 99, 115, 145, 152, 160, 299].

Structure	E, N/m	G, N/m	ν
α-graphyne	33.7	8.7	0.80
	42.8 [99]	8 [115]	0.72 [99]
β_1-graphyne	86.5	27.4	0.56
	93.6 [99]	25 [115]	0.52 [99]
γ_1-graphyne	154	61.2	0.37
	163 [99]	77 [9]	0.38 [99]
γ_1-graphdiyne	111	38.2	0.45
	118.6 [99]	41.6 [4]	0.40 [99]

Table 3.8. Extreme values of Young's modulus E (global maxima and minima E_{max}, E_{min}), Poisson's ratio ν (global maxima and minima ν_{max}, ν_{min}), shear modulus G (global maxima and minima G_{max}, G_{min}), and anisotropy coefficients $\Delta_1 \times 10^{-3}$, $\Delta_2 \times 10^{-3}$ for orthorhombic crystals (β_3-, γ_2-graphyne).

Structure	E_{min}, N/m	E_{max}, N/m	G_{min}, N/m	G_{max}, N/m	ν_{min}	ν_{max}
β_3	56.6	150	18.2	55.2	0.07	0.57
γ_2	103	321	59.2	85.0	0.13	0.65

Two configurations of graphyne (β_1, γ_1) and γ_1-graphdiyne present a Poisson's ratio very close to 0.5. Poisson's ratio equals 0.5 is a characteristic of an incompressible 3D material (like rubber) which conserves volume when subjected to axial strain. The same was observed previously for γ_1-graphyne [204].

Regarding the comparison between different graphynes, it can be mentioned that the increasing amount of acetylenic linkage causes a faster decay in the mechanical properties of graphyne. Therefore, α-graphyne is the weaker structure of the graphyne family.

The global extreme values of Young's modulus E, Poisson's ratio ν and shear modulus G for β_3-graphyne and γ_2-graphyne are given in Table 3.8. Orientation dependencies for Young's modulus E, Poisson's ratio ν and shear modulus G of β_3-graphyne and γ_2-graphyne are shown in Fig. 3.3. A higher anisotropy is found in β_3-graphyne ($E_{max}/E_{min} = 2.65$, $G_{max}/G_{min} = 3.03$) than in γ_2-graphyne ($E_{max}/E_{min} = 2.24$, $G_{max}/G_{min} = 1.44$).

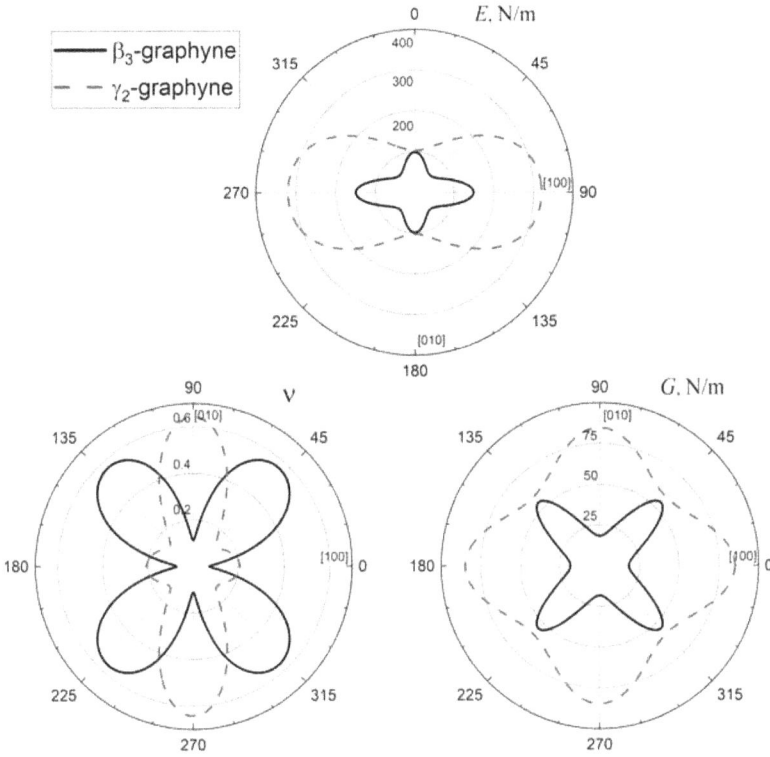

Fig. 3.3. Orientation dependencies of Young's modulus, Poisson's ratio and shear modulus for β_3-graphdiyne and γ_2-graphdiyne. Reprinted with permission from [212].

The analysis shows that Young's modulus and shear modulus for most graphynes are smaller than those for graphene and graphane. Only γ_2-graphyne has Young's modulus and shear modulus close to graphene and graphane. Poisson's ratio for graphynes turns out to be greater than for graphene and graphane. The highest value of Young's modulus and shear modulus is detected in γ_2-graphyne, and the lowest values in α-graphyne. And α-graphyne has the highest value of Poisson's ratio.

In both graphyne orientations, Young's modulus demonstrates a degrading trend with the increase of number of acetylene groups. This is reasonable because the in-plane honeycomb structure formed by hexagonal rings (graphene) is the stiffest known. Compared to the 320 N/m in-plane stiffness for graphene, the Young's modulus for the graphyne family is about 10–50% of that of graphene.

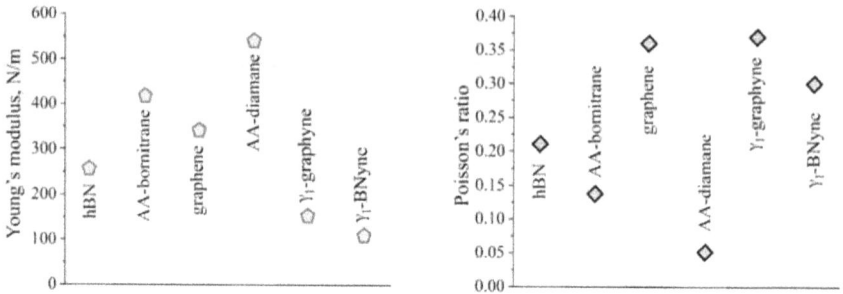

Fig. 3.4. Young's modulus (left) and Poisson's ratio (right) for different 2D structures with hexagonal anisotropy.

Figure 3.4 presents the Young's modulus and Poisson's ratio for different 2D structures with hexagonal anisotropy. As can be seen, 2D carbon polymorphs with different morphology have quite different mechanical characteristics. The Young's modulus is higher for structures with sp^3 hybridization than for sp^2. However, Poisson's ratio for bornitrane and diamane are the lowest. For one-atom-thick structures such as graphene, graphyne and BNyne, lower Young's moduli are found, while the Poisson's ratios are higher than 0.35.

3.1.4 Diamond-Like Phases

Fulleranes with cubic anisotropy are CA3, CA7, CA8, CA9 and CB. The elasticity of cubic crystals is characterized by three independent matrix compliance moduli: s_{11}, s_{12}, s_{44}. Tables 3.9 and 3.10 present compliance s_{ij} and stiffness c_{ij} coefficients, maximal and minimal values of Young's modulus and Poisson's ratio along different lattice directions for fulleranes with cubic anisotropy. The details of the calculations for the elastic constants of diamond-like phases (DLPs) can be found in [151, 232, 233].

Table 3.9. Compliance s_{ij} and stiffness c_{ij} coefficients for fulleranes with the cubic anisotropy.

Fullerane	s_{11}, TPa^{-1}	s_{44}, TPa^{-1}	s_{12}, TPa^{-1}	c_{11}, GPa	c_{44}, GPa	c_{12}, GPa
CA3	1.87	2.496	−0.44	625	401	192
CA7	8.12	3.64	−3.82	750	275	667
CA8	1.67	5.91	−0.299	650	169	142
CA9	3.73	7.31	−0.90	316	137	101
CB	5.53	10	−2.0	306	99.9	174

Table 3.10. Maximal and minimal values of Young's modulus and Poisson's ratio along different directions for fulleranes with cubic anisotropy.

Fullerane	E_{min}, GPa	E_{max}, GPa	$\nu_{[100],[001]}$	$\nu_{[001],[110]}$	$\nu_{[1\bar{1}0],[110]}$	$\nu_{(111),[111]}$
CA3	535	**860**	0.24	0.33	−0.07	0.07
CA7	123	**644**	0.47	1.14	−0.41	0.33
CA8	**599**	430	0.18	0.14	0.37	0.27
CA9	268	**325**	0.24	0.28	0.13	0.19
CB	181	**260**	0.36	0.47	0.17	0.30

Of the five cubic fulleranes, only one (CA8) has negative anisotropy, and the maximum Young's modulus will be observed in the direction [100], and the minimum in the direction [111]. Among the fulleranes, the CA3 phase, which has positive anisotropy, demonstrates the largest Young's modulus.

Table 3.10 shows that there are two partial auxetics among fulleranes: CA3 and CA7 phases. Both have a negative Poisson's ratio for the orientation [110], [110]. No complete auxetics were found among the DLPs. The lowest Poisson's ratio is found for the equilibrium phase CA7, while the highest Poisson's ratio is observed along [001], [110] and is equal to 1.14. The average Poisson's ratio for all the presented phases is positive, in the range from 0.15 to 0.42.

Among DLPs, TA1, TA3, TA5, TA6 and LA3 belong to the tetragonal anisotropy, TB and CA2 to the hexagonal, and TA8 to the trigonal. Table 3.11 shows the compliance and stiffness coefficient for these DLPs.

Figure 3.5 shows the orientation dependence of the Young's modulus for CA2, TA1, TA3, TA5, TA6, TA8 and TB. The maximum value of 521 GPa is achieved for CA2 under tension in the [001] direction. The minimum value of the Young's modulus for CA2 is $E_{min} = 181$ GPa. As can be seen from Fig. 3.5a, the Young's modulus is orientation-dependent and in [001] direction significantly exceeds those in other directions. For TB, the maximum Young's modulus is 1730 GPa (see Fig. 3.5b) and is observed in the direction of extension at an angle of 38° to [0001]. A high Young's modulus (983 GPa) is also observed under extension along the [0001] direction. The lowest value of Young's modulus (573 GPa) corresponds to the crystallographic direction [2110].

To study the variability of Young's modulus the (100), (010), and (001) planes were considered. The Young's modulus are $E_{max} = 1043$ GPa, $E_{min} = 504$ GPa, $E_{max}/E_{min} = 2.07$ GPa, $E_{[100]} = 504$ GPa, $E_{[010]} = 504$ GPa and $E_{[001]} = 1043$ GPa. Figure 3.5c shows the orientation

Table 3.11. Compliance s_{ij} and stiffness c_{ij} coefficients for DLPs with tetragonal, hexagonal and trigonal anisotropy.

DLP	s_{11}, TPa^{-1}	s_{12}, TPa^{-1}	s_{13}, TPa^{-1}	s_{33}, TPa^{-1}	s_{44}, TPa^{-1}	s_{66}, TPa^{-1}	s_{14}, TPa^{-1}
CA2	2.51	0.09	−0.36	1.92	18.4	−	−
TA1	2.15	−0.88	−0.52	2.3	7.90	5.51	−
TA3	1.9	−0.71	−0.52	2.58	7.76	4.96	−
TA5	1.55	−0.49	−0.30	1.14	5.60	2.16	−
TA6	1.47	−0.01	−0.126	0.82	4.73	2.77	−
TA8	1.98	−0.94	−0.06	0.96	3.88	−	−0.005
TB	1.656	−0.159	−0.29	0.999	10.174	−	−
LA3	2.03	−2.03	−0.04	8.15	11.47	3.44	−

DLP	c_{11}, GPa	c_{12}, GPa	c_{13}, GPa	c_{33}, GPa	c_{44}, GPa	c_{66}, GPa	c_{14}, GPa
CA2	413	−1.63	87.3	555	161	−	−
TA1	652	318	196	461	130	182	−
TA3	706	294	165	463	129	201	−
TA5	820	350	226	989	179	463	−
TA6	1854	55	−8.59	1214	442	221	−
TA8	657	315	67	1051	257	−	0.5
TB	600	−30.5	155	1067	3614	−	−
LA3	626	79	40	1232	87	290	−

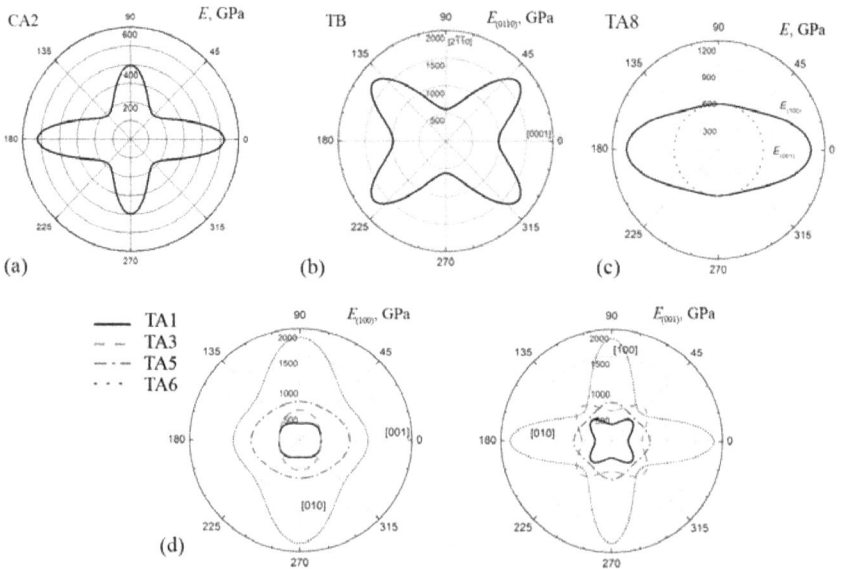

Fig. 3.5. Orientation dependence of Young's modulus for (a) CA2, (b) TB, (c) TA8, and (d) Young's modulus $E_{(100)}$ (left) and $E_{(001)}$ (right) for four tubulanes with the same anisotropy TA1, TA3, TA5 and TA6.

dependence of Young's modulus for two planes (100) and (001). Since the compliance coefficient s_{14} is significantly smaller than the others, the behavior of Young's modulus for the planes (100) and (010) is similar. The maximum value of Young's modulus is 1043 GPa and corresponds to the [001] direction. The minimum value of Young's modulus (504 GPa) is achieved in the (001) plane.

Tetragonal phases TA1, TA3, TA5, TA6, LA3. To study the variability of Young's modulus and the shear modulus of tubulanes, the (100) and (001) planes were chosen. The behavior of the elastic characteristics in the (010) and (100) planes is similar. Figure 3.5d shows the orientation dependence of Young's modulus for the TA1, TA3, TA5, and TA6 phases in the (100) and (001) planes. As can be seen, TA1 has the lowest values of the Young's modulus compared to TA3, TA5 and TA6. The maximum value of the Young's modulus for TA1 is revealed during tension in the (001) plane at an angle of 45°. The maximum value of Young's modulus is also observed in this direction for TA3. Covalent bonds lie along this direction for the TA1 and TA3 phases. For TA5 and TA6, the maximum Young's modulus was found in [100] and [010] directions.

For TA6 and TB, the maximum Young's modulus exceeds 1 TPa, which is greater than the Young's modulus of graphite under tension along the graphene planes. For TA6, the Young's modulus in the (100) plane is always greater than 1 TPa. The TA6 phase has the greatest variability of Young's modulus ($E_{max}/E_{min} = 2.58$) among all the considered tubulanes with tetragonal anisotropy. For TB, the value of the Young's modulus is grater than 1 TPa.

Table 3.12 presents Poisson's ratio for TA1, TA3, TA5 and TA6. Analysis of the variability of the Poisson's ratio for the TA1, TA3, TA5 and TA6 phases showed that TA6 is a partial auxetic. The minimum value of the Poisson's ratio for TA6 is 0.01. This phase also has the largest difference $\nu_{max} - \nu_{min} = 0.63$ between the extreme values. The other three phases (TA1, TA3, TA5) have a positive Poisson's ratio for any orientation.

Table 3.12. Maximal (ν_{max}), minimal (ν_{max}), average ($\langle \nu \rangle$), and oriented values of the Poisson's ratio.

DLP	ν_{min}	ν_{max}	$\langle \nu \rangle$	$\nu_{[010],[100]}$	$\nu_{[001],[100]}$	$\nu_{[100],[010]}$	$\nu_{[001],[010]}$	$\nu_{[100],[001]}$	$\nu_{[010],[001]}$
TA1	0.03	0.58	0.25	0.58	0.13	0.58	0.13	0.16	0.16
TA3	0.1	0.42	0.22	0.42	0.2	0.42	0.2	0.14	0.14
TA5	0.15	0.57	0.31	0.48	0.16	0.48	0.16	0.21	0.21
TA6	−0.01	0.62	0.17	0.03	−0.007	0.03	−0.007	−0.005	−0.005
TA8	0.03	0.48	0.23	0.47	0.03	0.47	0.03	0.07	0.07

The minimum difference between the extreme values is observed for TA3 $(\nu_{max} - \nu_{min} = 0.32)$. The maximum values of the Poisson's ratio for TA1 and TA5 exceed 0.5, which corresponds to the upper limit for isotropic material. Under tension along [001] direction, Poisson's ratio remains constant. The Poisson's ratio $\nu[001],[100]$ for the TA1 and TA3 phases corresponding to transverse strain in the [001] direction is smaller than that in the [010] direction. The Poisson's ratio for TA6 is changing slightly near zero.

Analysis of the Poisson ratio for TA8 shows that this tubulane is not auxetic (see Table 3.12). The variability of the Poisson ratio under tension in the [100], [010], and [001] directions showed that for tension in the [100] direction, the Poisson ratio can reach both the maximum (0.47) and minimum (0.03) values. It is worth noting that the orientation dependences under tension in the [100] and [010] directions are similar, since the compliance coefficient of the TA8 phase is significantly less than those of the other phases.

Among the graphene-based DLPs, only two stable phases, LA3 and LA6, were found. The LA3 phase has a tetragonal syngony, and the LA6 phase has a rhombic syngony. Since both phases are considered in similar planes, it seems rational to describe them together and compare the obtained curves. Elastic constants for LA3 are presented in Table 3.11.

There are nine compliance constants for LA6: $s_{11} = 2.45$ TPa^{-1}, $s_{12} = -1.4$ TPa^{-1}, $s_{13} = -0.35$ TPa^{-1}, $s_{22} = 2.23$ TPa^{-1}, $s_{23} = -0.13$ TPa^{-1}, $s_{33} = 1.21$ TPa^{-1}, $s_{44} = 2.15$ TPa^{-1}, $s_{55} = 2.36$ TPa^{-1}, $s_{66} = 2.52$ TPa^{-1}. And nine corresponding stiffness constants for LA6: $c_{11} = 720$ GPa, $c_{12} = 473$ GPa, $c_{13} = 271$ GPa, $c_{22} = 761$ GPa, $c_{23} = 229$ GPa, $c_{33} = 933$ GPa, $c_{44} = 454$ GPa, $c_{55} = 423$ GPa, $c_{66} = 395$ GPa.

Figure 3.6 presents the orientation dependence of the Young's modulus for LA3 and LA6. For LA6 under tension along (100) maximal Young's modulus is $E_{max} = 838$ GPa can be found at 21.2°. For (010) plane, maximal Young's modulus is 861 GPa at 22.9°. For tension along (001) plane there is a great difference between the values of the Young's modulus $E_{max}/E_{min} = 2.55$, $E_{min} = 1039$ GPa.

For LA3, the orientation dependencies in the (100) and (010) planes have the same shape (see Figure 3.6) because of its tetragonal anisotropy. The maximum value of Young's modulus in these planes is 1235 GPa and is observed in the [001] direction. Under tension along the [100] and [010] directions, Young's modulus is two times smaller than in the [001] direction $(E_{[100]} = E_{[010]} = 671$ GPa). The orientation dependencies in the (100)

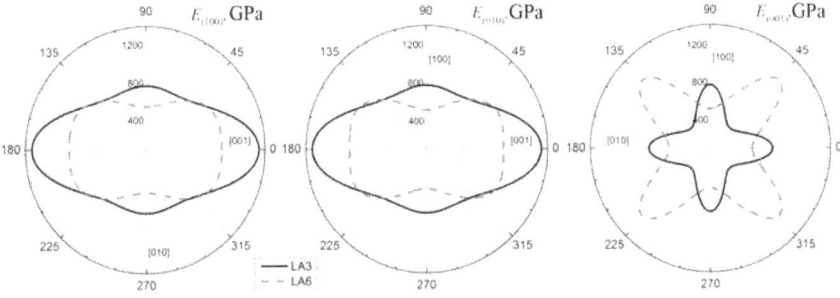

Fig. 3.6. Orientation dependence of Young's modulus for LA3 and LA6. Reprinted with permission from [233].

Table 3.13. Maximal, minimal and oriented Poisson's ratio for LA3 and LA6.

DLP	ν_{min}	ν_{max}	$\nu_{[010],[100]}$	$\nu_{[001],[100]}$	$\nu_{[\bar{1}00],[010]}$	$\nu_{[001],[010]}$	$\nu_{[100],[001]}$	$\nu_{[010],[001]}$
LA3	−0,13	0.51	−0.13	0.03	−0.13	0.06	0.06	0.06
LA6	−0.14	0.66	0.57	0.14	0.63	0.06	0.29	0.11

and (010) planes is similar for LA6. The values of the Young's modulus for tension along [100] and [010] directions differ slightly ($E_{[100]} = 407$ GPa and $E_{[010]} = 447$ GPa). Young's modulus $E_{[010]}$ is greater due to the fact that in the [010] direction there is an additional bond between carbon atoms. The maximum Young's modulus for the LA6 carbon phase is found during tension in the (001) plane at an angle close to 45° to the [100] direction. The minimum value of the Young's modulus for LA3 phase is found in the same direction. For LA6, the minimum value of the Young's modulus is found in the [100] direction. Note that the maximum Young's modulus for LA3 (1235 GPa) and LA6 (1039 GPa) are greater than 1000 GPa observed for graphene.

The analysis of the variability of the Poisson's ratio showed that the LA3 and LA6 phases are partial auxetics (see Table 3.13). For LA3 and LA6, the minimal Poisson's ratio are close ($\nu_{min} = -0.13$ for LA3 and $\nu_{min} = -0.14$ for LA6). The maximum values of the Poisson's ratio for LA3 and LA6 exceed 0.5, which is the upper limit for isotropic material. The largest difference between the extreme values is revealed for LA6: $\nu_{max} - \nu_{min} = 0.8$. Under tension along [100], LA3 exhibits a negative Poisson's ratio.

The mechanical properties of DLPs depend on the hybridization of C atoms and change even at low concentration of sp^3 hybrid atoms in disordered graphite. In case of DLPs with almost complete sp^3 hybridization, the elastic constants can vary within wide limits depending on the type of translation cell. Figure 3.7 presents an overview of the Young's modulus and Poisson's ratio.

For anisotropic structures, Young's modulus and Poisson's ratio, depending on the tensile direction with respect to the crystallographic axis, can considerably change over different directions. Thus, Figure 3.7 presents only maximal and minimal values. The maximal values of Young's modulus for some DLPs (TA6, TA8, TB, and LA3) are even bigger than for diamond,

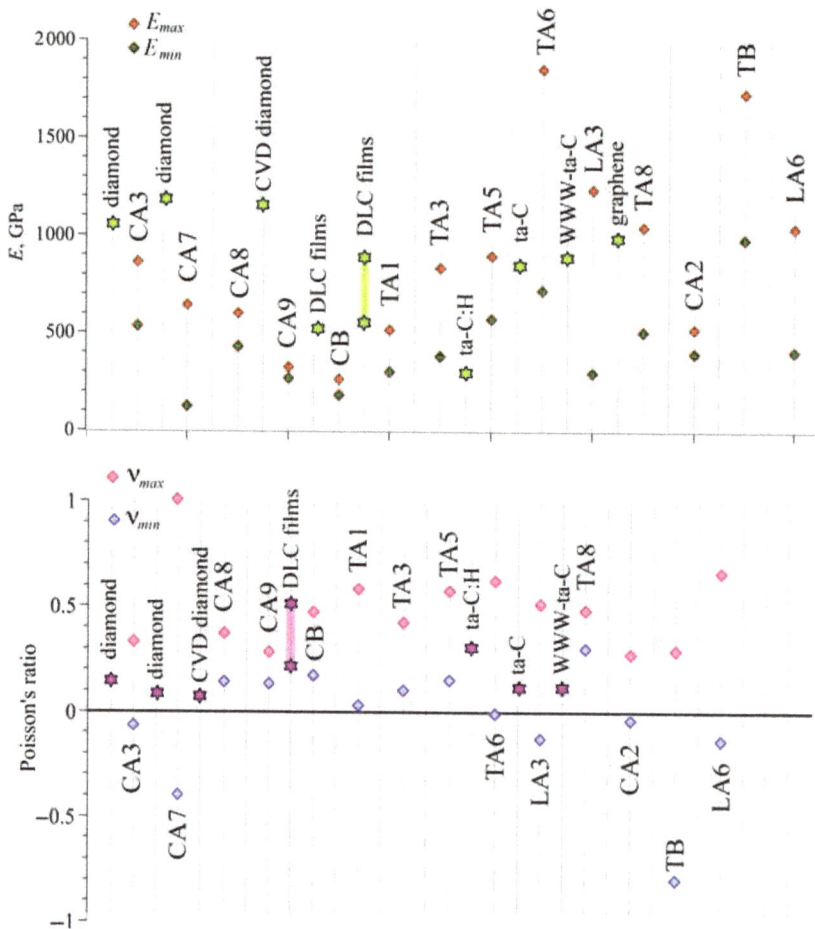

Fig. 3.7. Young's modulus and Poisson's ratio for all studied DLPs (shown by rhombus) compared with the literature (shown by stars). Reprinted with permission from [12].

which is 1144.6 GPa, and for CVD diamond 1143 GPa. However, for DLP films, Young's modulus is almost two times lower: 500–530 GPa. Very well-known DLPs such as ta-C:H, ta-C, and WWW-ta-C have a Young's modulus of 300, 757, and 829 GPa, respectively.

The lowest Young's modulus is found for cubic C9 and CB phases. Phase LA3 possesses a considerable difference between maximal and minimal values of Young's modulus, about four times. It is very similar to graphite, which exhibits almost zero Young's modulus along the [001] direction and a maximal Young's modulus of 1000 GPa along the [010] direction [94]. The reason for low strength for DLPs is the distribution of lattice bonds. If one of the important bonds is aligned along the tensile direction or angle rotation is blocked, the Young's modulus will be lower. DLPs CA7, TA3, TA6, LA3, TA8, and LA6 are highly anisotropic. For CA7 and LA3, the difference between minimal and maximal Young's modulus is greater than four. The best isotropy was found for CA2, CA9, and CB.

Considerable difference between the values of Poisson's ratio from −0.8 to +0.66 can be seen, while for some DLPs (CA3, TA1, TA6, and CA2) minimal Poisson's ratio is very close to zero. DLPs CA3, CA7, TA6, LA3, CA2, TB, and LA6 possess negative Poisson's ratio. Some DLPs are partially auxetic since they possess negative Poisson's ratio in some directions, and some are non-auxetic at all. There are no full auxetics among DLPs. Tubulane TA6 is a partial auxetic and possesses a considerable difference in Poisson's ratio along different directions. The minimal Poisson's ratio for TA6 is −0.01. For tubulane TA6, the biggest difference between maximal and minimal values was found, while average Poisson's ratio for TA6 is positive and equal to 0.17. For tubulane TA6, maximal Young's modulus is observed in the [100] and [010] directions.

3.2 Deformation Behavior and Tensile Strength

Recently, numerous research works and reviews have been devoted to the study of the mechanical properties of 2D materials, both in experiment and simulation. For such nanostructured materials, the measurement technique is highly important, as well as the purity of graphene, presence of defects or additional layers. Experimental methods such as AFM nanoindentation, micro-/nano-mechanical devices, and pressurized bulge testing to date have been successfully used to visualize and estimate the mechanical properties of graphene and other 2D materials. However, in most of the experiments, deformation of 2D structure is almost uncontrollable and considerable efforts have been devoted to developing experimental approaches for obtaining controllable analysis of the mechanical properties. A full description of different experimental methods was presented in a review work [60], while in [266] some simulation techniques were described.

In a number of theoretical and experimental works, the appearance of intrinsic ripples have been observed for graphene deposited on the substrates, and shown to be necessary to stabilize the suspended graphene against the thermal instabilities for 2D systems. However, it was also experimentally demonstrated that the rippling can be eventually fully suppressed by depositing 2D material onto an appropriate substrate.

Considering the deformation behaviour, the applied methodology should be taken into account. Also, the edge orientation of graphene (armchair or zigzag) is very important, as well as temperature and presence of defects.

3.2.1 *Graphene*

Deformation behaviour of graphene has been numerously discussed in literature [37, 60, 266]; however, the majority of works considered planar graphene. The experimental analysis of the fracture strength of graphene (and other 2D structures) is complicated and cannot be directly conducted as for conventional 3D materials; however, the results obtained in different simulations can considerably differ. Figure 3.8 presents the overview of tensile strength and fracture strain of graphene obtained from simulations [266].

As can be seen from Fig. 3.8a, ultimate tensile strength (UTS) of graphene is commonly defined in the range from 90 to 140 GPa, with lower

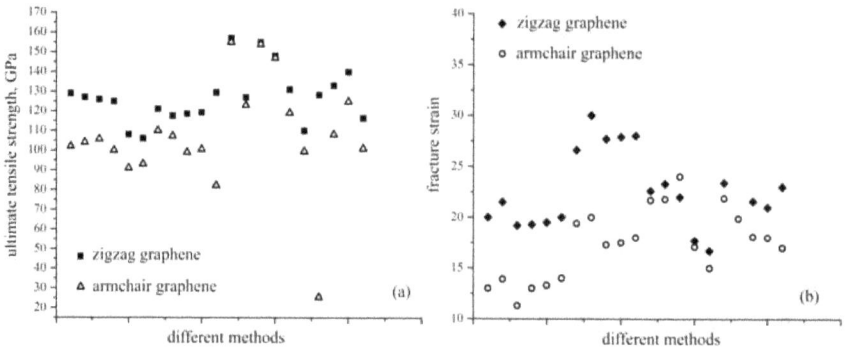

Fig. 3.8. Tensile strength and fracture strain of graphene obtained from simulation.

UTS for armchair graphene. Fracture strain (Fig. 3.8b) is in the range from 10–30%, lower for armchair direction. Higher values of UTS and fracture strain were obtained for lower temperatures.

The other important issue is the effect of lattice defects on graphene strength, which was partially discussed in Chapter 2. Figure 3.9 presents the stability region of flat graphene: defect-free (gray curve) and with Stone-Wales (SW) defect (black curve). In this case, rippling is forbidden during tension. As can be seen, there is no difference in the fracture strain for defect-free graphene and graphene with SW under tension along the zigzag direction.

Let us here present the deformation behavior of graphene taking into account the possibility of graphene rippling. Ripple formation is well known for graphene and can considerably affect their properties. It is well known that graphene is a high-strength material with enormously large Young's modulus [145]; however, different simulation techniques give very different results, as well as experiments [8, 23, 37, 56, 133]. Under dynamic loading conditions, graphene exhibits brittle fracture instead of ductile due to insufficient structural relaxation [306]. The deformation energy in

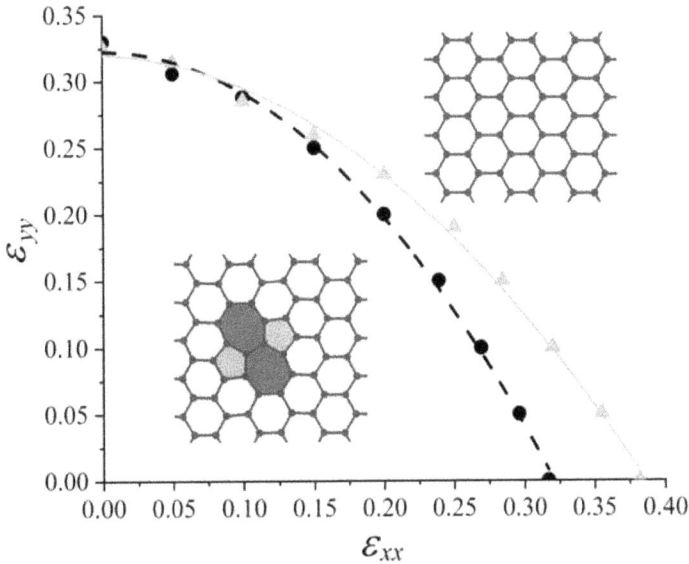

Fig. 3.9. Stability region of flat graphene: defect-free (gray curve) and with SW defect (black curve).

free-standing graphene can be released by the formation of out-of-plane wrinkles [303]. The tensile strain along one direction and the compressive strain along the other direction lead to graphene instability and wrinkling [14, 15]. However, wrinkling significantly affects the elastic modulus and fracture strain and strength [91, 218]: even for small-amplitude corrugations of graphene, the fracture strength and elastic modulus decrease.

Figure 3.10 shows the stress-strain curves during tension at 0 K along the armchair (a) and zigzag (b) directions for defect-free graphene and graphene with the SW defect and representative dislocation dipole (DD_4 and DD_{10}) when wrinkling is allowed. Here 4 and 10 are the distance between dislocations in number of hexagons: DD_4 (dipole shoulder $l = 11$ Å) and DD_{10} (dipole shoulder $l = 25$ Å).

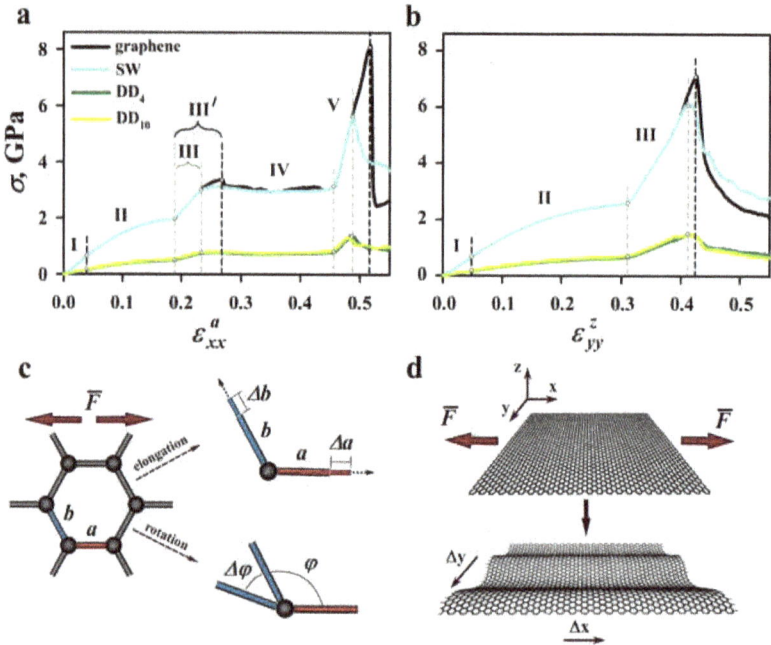

Fig. 3.10. Stress-strain curves for defect-free graphene, graphene with SW, DD_4 and DD_{10} under tension along the (a) armchair and (b) zigzag direction at $T = 0$ K. Dashed lines define different deformation regions numbered as I–V: black for graphene and green for graphene with defects. (c,d) Illustration of the deformation mechanism for graphene under tensile loading. The two deformation modes are rotation/elongation and wrinkling. Here, F represents the tensile force, a and b represent the bond lengths and φ is the in-plane angle between two bonds. Reprinted with permission from [2].

For tension along the armchair direction (Fig. 3.10a), five deformation stages can be defined: I – linear elastic region, II, III – strain-hardening regions, IV – steady state region, V – pre-critical strain. For tension along zigzag, there are three regions: I – linear elastic region, II – strain-hardening region, and III – pre-critical strain. The stress-strain curves in Fig. 3.10a,b are divided by dashed lines for clarity: black for graphene and green for graphene with defects.

Figure 3.10c,d presents two main deformation mechanisms, which are the elongation and rotation of bonds and graphene wrinkling. Note that all these mechanisms are not contradictory, but act simultaneously, resulting in a rather high fracture strain, but low strength. Here, F represents the tensile force acting on graphene, a and b represent the bond lengths, and φ is the in-plane angle between two adjacent bonds.

For tension along armchair (black curve, Fig. 3.10a), at hardening region II an irreversible elongation of bond a was observed with rotation of bond b. When the rotation of the valence angles is limited, further elongation of bonds a and b occurs at the beginning of the next strain-hardening region III. Further (region V), rapid elongation of bond a and subsequent fracture occured.

For tension along zigzag (Fig. 3.10b), the hardening during region II is defined by the simultaneous rotation and elongation of the bond b. In the next pre-critical region III, the length of the bond b reaches its critical value, leading to stress growth and subsequent fracture. The presence of the SW defect reduces the fracture strength and strain.

Under tension of defect-free graphene, the nucleation of multiple cracks starts at a random location and is followed by crack propagation. For armchair graphene, crack grows incrementally at an angle of 45° to the tensile direction.

For graphene with dislocation dipoles, the same deformation mechanisms were found, but the steady state region is longer and fracture strength is much lower. The difference in the dipole arm has almost no effect on the graphene strength.

3.2.2 *Graphyne*

The strain energy and deformation of graphene can be well described by continuum elasticity theory, which can also be applied to graphyne. It can be assumed that under small deformation graphyne can be considered as a linear elastic 2D layer. Again, as for graphene, different types of loading conditions can be simulated. Two main orientations of graphynes can be distinguished – armchair and zigzag (see Fig. 3.11). In the present subsection, two cases are considered: graphynes with and without rippling.

Tensile strength and Young's modulus can be different for different graphyne morphology [296]. As can be seen, graphene demonstrates the highest UTS, as well as the highest fracture strain at tension in both directions. For armchair direction, tensile strength is different for different number of acetylenic bonds, and much lower than for graphene itself, with fracture strain not affected. For zigzag direction, UTS is the same for all the considered structures, but fracture strain is different, which is explained by the difference of deformation mechanisms along armchair and zigzag directions. The best way to analyze the deformation behavior of carbon nanostructures is to take into account the changes of the covalent

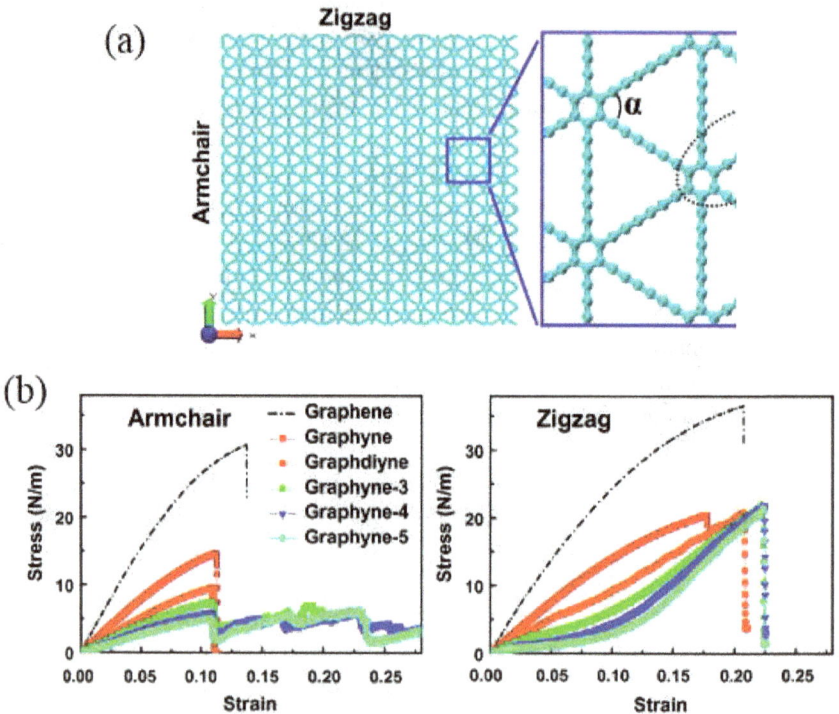

Fig. 3.11. (a) Snapshot of graphdiyne, along with the definitions of edges. (b) Stress–strain relations of the graphyne family under tension. Reprinted with permission from [296].

bonds and angles. For zigzag direction, increase of the number of acetylenic bonds results in the difference of the valence angles in the structure and subsequently in the increase of fracture strain.

Not only the difference of the number of acetylenic bonds but also the type of graphyne (α, β or γ) gives different values of strength. For example, for γ_1-graphyne, Young's modulus under tension along armchair edge was 532.5 GPa (or 170.4 N/m without consideration of the sheet thickness), UTS was 48.2 GPa and maximum strain was 0.0819 [52]. For tension along zigzag edge, Young's modulus was 700.0 GPa (or 224.0 N/m). In this case, the elastic modulus increased to 888.4 GPa (284.3 N/m), which is explained by the alignment of the acetylenic groups towards the applied strain. The UTS along zigzag was 107.5 GPa and strain was 0.1324.

As for graphene, wrinkling of graphynes considerably decrease their strength. Figure 3.12 shows the UTS and fracture strain for graphene, four graphynes, and graphdiyne during uniaxial tension along zigzag and armchair edges. The graphynes and γ_1-graphdiyne have close values of the UTS and fracture strain; however, the higher the graphyne density, the higher the UTS. For γ_2-graphyne, considerable differences in UTS between zigzag and armchair are observed. Interestingly, fracture strain of graphene is close to that of graphynes, which is the result of graphene corrugation.

For all the considered structures, the same deformation stages can be defined: (1) elastic strain up to about $\varepsilon = 0.05$; (2) structural transformation from flat to wrinkled structure until about $\varepsilon = 0.2$ with the simultaneous changes of covalent bonds and angles; (3) flattening of the wrinkles with sharp increase of covalent bonds; and (4) pre-critical regime when the structure was flattened again and highly stressed with the sharp changes of valence angles.

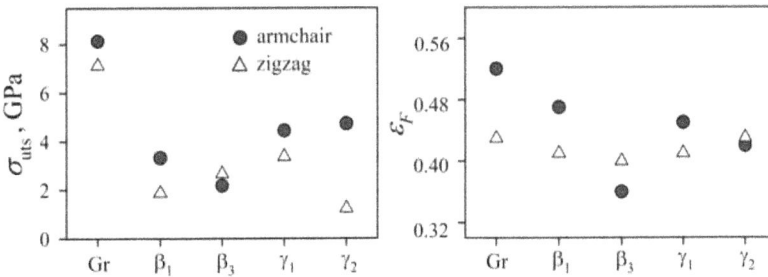

Fig. 3.12. UTS and fracture strain for graphene and graphynes under uniaxial tension.

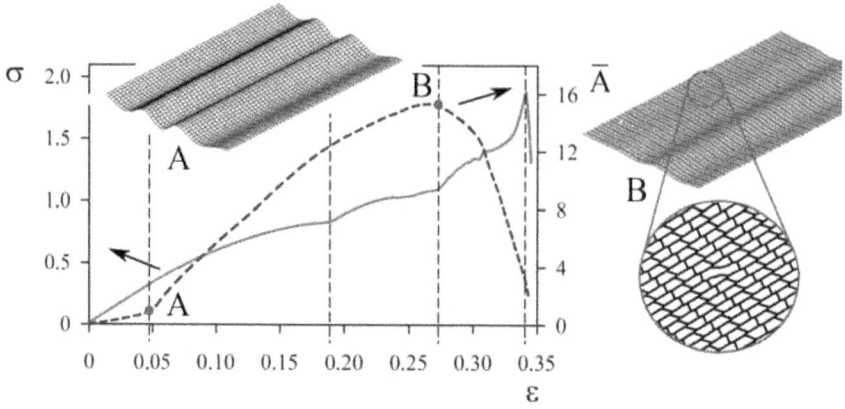

Fig. 3.13. Stress (σ, left axis) and ripple amplitude ($\overline{A}$, right axis) as a function of strain for β_3-graphyne under tension. Snapshots of the structure are presented at points A and B. Stress is in GPa, ripple amplitude in Å.

Figure 3.13 shows an example of the stress-strain curve for β_3-graphyne under tension and snapshots of the structure at critical points. Ripple amplitude is also presented as a function of strain. During linear elastic deformation, only slight elongation of the bonds and change of the valence angles occurred. In the second and third stages, the simultaneous action of two deformation mechanisms took place: the wrinkle amplitude increased, and bond elongation with changes of valence angles took place. At the last deformation stage, covalent bonds cannot further elongate and tensile deformation was realized from flattening and increase of the valence angles.

3.2.3 *Diamane*

The atomic lattice of diamane is six-fold-rotation symmetric, and the in-plane orientation can be described by a chiral angle θ: $0° \leq \theta \leq 30°$, where $0°$, $30°$ correspond to the armchair and zigzag directions. In [215], the orientation dependence of the diamane strength was investigated. Figure 3.14a presents the stress-strain curves for D-AA+H during uniaxial tension for diamane with different chiral angles: $0°$ (purple curve), $10°$ (blue curve), $15°$ (green curve), $20°$ (red curve), $30°$ (black curve). The diamane morphology (AB or AA) has almost no effect on the mechanical properties. Figure 3.14b,c present the ultimate tensile strength (σ_{UTS}) (b) and the fracture strain (ε_F) (c) as a function of chiral angle for both AB and AA morphologies for comparison. It was found that the interlayer

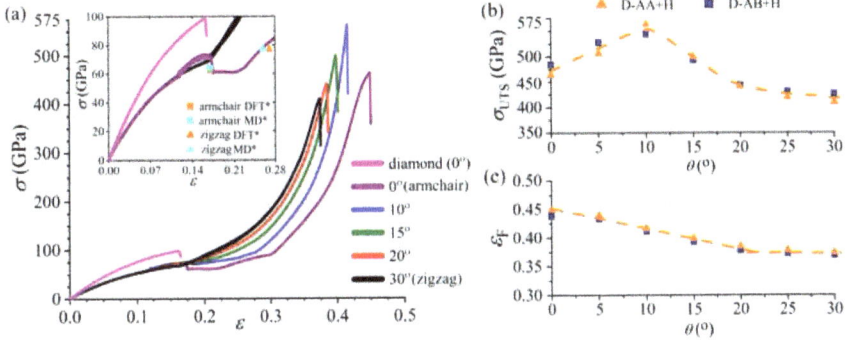

Fig. 3.14. (a) Stress-strain curves for D-AA+H during uniaxial tension for diamane with different chiral angles. The inset presents the ultimate tensile strength for diamane obtained by DFT [177] and MD [285]. (b,c) Ultimate tensile strength (σ_{UTS}) (b) and fracture strain (ε_F) (c) as a function of chiral angle for both AB and AA morphology. Reprinted with permission from [215].

covalent bonds remain almost unchanged during tension, which explains the similarity between the tensile behavior of D-AB and D-AA.

Previously [285], the UTS and fracture strain calculated by MD for tension along armchair direction were respectively found to be 64 GPa and 0.17 and along zigzag direction 77 GPa and 0.26, which agree with DFT results [177]. In [215], UTS and fracture strain are much larger and explained by the phase transition occurring during tensile loading (inset to Fig. 3.14a). Figure 3.14b shows that σ_{UTS} increases with the increase of the chiral angle of diamane up to 10°, and then decreases. The maximum σ_{UTS} for diamane was 562 GPa for D-AA+H, and 543 GPa for D-AB+H. A decrease in the fracture strain (see Fig. 3.14c) was observed for chiral angles from 10° to 20°, which has also been previously found for graphene [62]. However, a significant increase in tensile strength and fracture strain was observed for graphene with chiral angles greater than 20°. The maximum fracture strain for D-AA+H (D-AB+H) was 0.449 (0.440) at the chiral angle of 0° (along the armchair direction). It can be seen that there is a relationship between the chiral angle of diamane and its mechanical properties, similar to graphene [50, 62].

3.2.4   *3D Graphenes*

3D carbon structures can be obtained based on various carbon polymorphs, such as graphene flakes, fullerenes, and carbon nanotubes. Figure 3.15

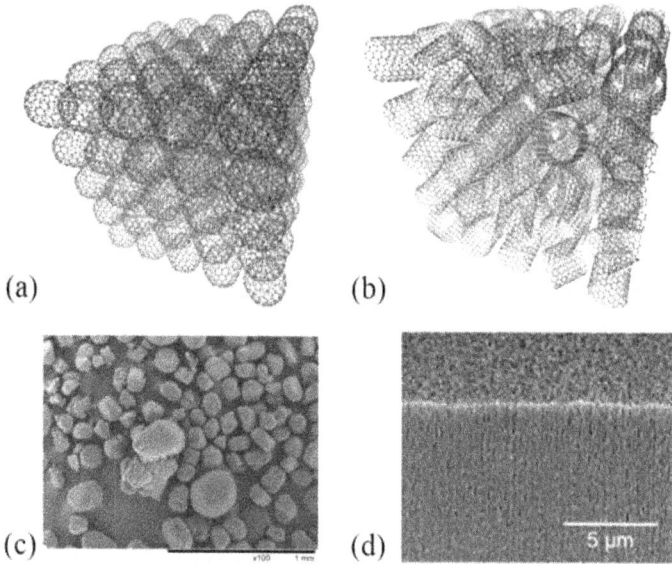

(a)   (b)

(c)   (d)

Fig. 3.15. (a) 3D structures with simple cubic packing: (a) system of fullerenes; (b) system of short CNTs in a model representation; (c) Coarse-crystalline powder of fullerite C_{60} in a scanning electron microscope; (d) CNT bundle.

presents examples of such structures (crumpled graphene is presented in Fig. 2.30). Deformation behaviour of these structures can be considered separately or in comparison.

3.2.4.1   *Crumpled Graphene, Fullerite and Carbon Nanotube Bundles Under Compression*

Figure 3.16 shows the pressure as a function of density for mono- (dashed line) and polydisperse (solid line) crumpled graphene and the initial structures. Crumpled graphene consists of differently oriented and crumpled graphene flakes, but for monodisperse the graphene flakes are of the same size. As can be seen, the polydisperse material is more easily deformed, since larger graphene flakes can undergo a greater degree of crumpling and can be easily deformed.

Figure 3.17 presents the pressure-density and stress-density curves for polydisperse crumpled graphene under hydrostatic and uniaxial (compression), together with the snapshots of the structural unit composed of five different graphene flakes. Here, pressure $p = (\sigma_{xx} + \sigma_{yy} + \sigma_{zz})/3$

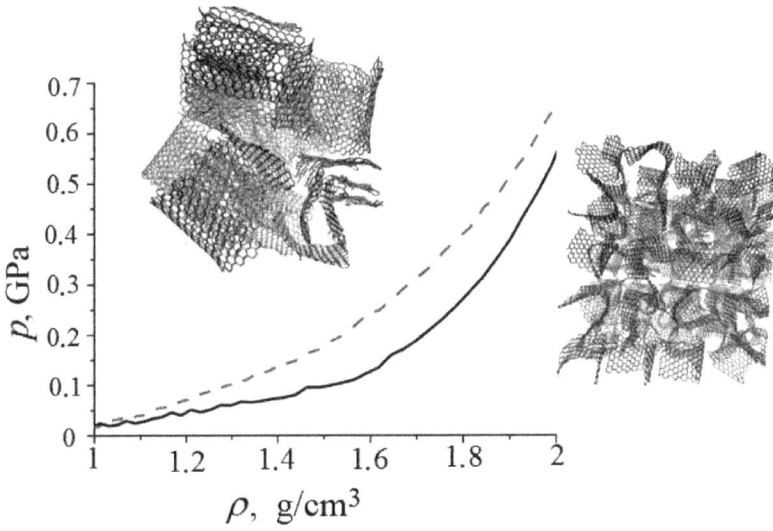

Fig. 3.16. Hydrostatic pressure as a function of density for the poly- (solid curve) and mono-disperse (dashed curve) structure. Initial structure of two types of crumpled graphene are also presented.

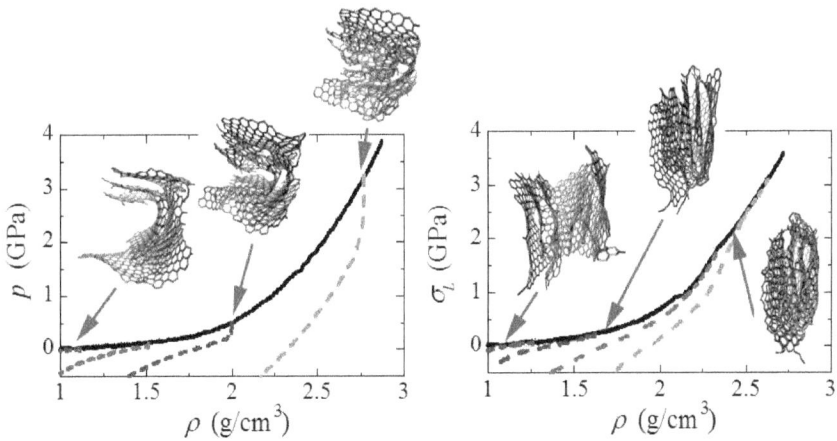

Fig. 3.17. Longitudinal stress (left) and hydrostatic pressure (right) as a function of density for the polydisperse crumpled graphene. Dashed lines show the pressure-density curves during unloading. Snapshots of the structural unit composed of five different graphene flakes during compression are depicted.

and longitudinal stress $\sigma_L = \sigma_{xx}$ for uniaxial compression along one (x) axis. Dashed lines show the pressure-density curves during unloading. The elasticity limits are $\rho = 1.25 \ \text{g/cm}^3$ for hydrostatic compression and $\rho = 1.5 \ \text{g/cm}^3$ for uniaxial compression. During uniaxial compression, graphene flakes are mostly bent with the formation of one large fold. During hydrostatic compression, considerable folding occurred with the formation of defects such as corners of double-folded sheets with a severely damaged structure.

Crumpled graphene is a non-Hookean medium, showing a nonlinear stress-strain relationship even at low strain values. The stiffness of crumpled graphene increases with increasing density due to the appearance of new van der Waals forces between graphene flakes and new covalent bonds at the flake edges.

Figure 3.18a presents the pressure-density curves under hydrostatic compression for three different 3D graphenes: crumpled graphene (G), fullerite (F) composed of fullerenes C_{240} and the system of carbon

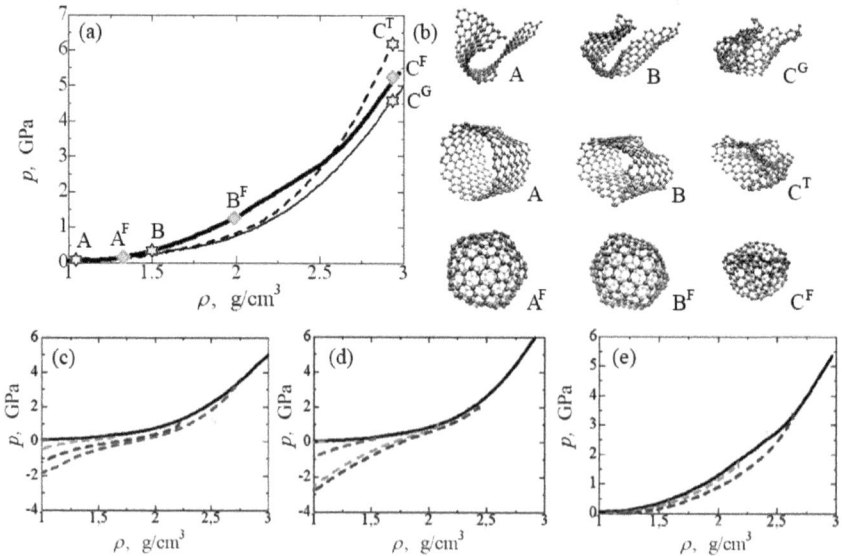

Fig. 3.18. (a) Hydrostatic pressure as a function of density for crumpled graphene (G - thin solid line), fullerite (F - thick solid line) composed of fullerenes C_{240} and the system of CNTs (T - dashed line). (b) Snapshots of the structural unit during compression. (c-e) Loading (solid lines) and unloading (dashed lines) pressure-density curves: (c) crumpled graphene; (d) system of CNTs; (e) fullerite.

nanotubes, CNTs (T). Figure 3.18b presents one structural unit during compression at critical points A, B, C. Figure 3.18c-e presents the loading (solid lines) and unloading (dashed lines) pressure-density curves. Initial structures are presented in Figs. 3.15 and 2.30.

As can be seen, fullerite demonstrates linear dependence between p and ε up to 0.08, while crumpled graphene and the system of CNTs are the non-Hookean medium and show nonlinear dependence $p \approx \varepsilon^2$ for low strain. Fullerite possesses maximal strength under compression up to a density of 2.5 g/cm^3, then the system of CNTs becomes strongest among the considered structures. This difference is due to the structure peculiarities. It is seen from the snapshots of the structural element that graphene flakes can be easily bent and deformed, while fullerene demonstrates considerable resistance to compression and remains an almost ideal form up to a density of 2.0 g/cm^3, which explains the highest strength of fullerite at initial deformation stages. However, at severe deformation, CNTs are collapsed and form a stable configuration, so this material becomes strongest at density above 2.5 g/cm^3. The collapse of fullerenes occurs instantly at sufficiently high pressures, with possible transformation into amorphous carbon.

Figure 3.19 presents the pressure-density curves for three fullerites: C$_{60}$ with simple cubic (SC) lattice, C$_{60}$ with face centered cubic (FCC) lattice and C$_{48}$ with SC lattice. Fullerite can be compressed to fairly high densities, close to the density of diamond. The deformation behavior of structures

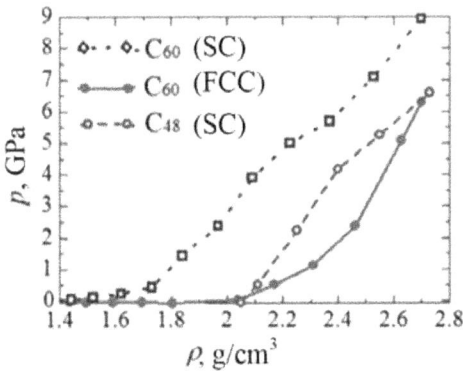

Fig. 3.19. Pressure-density curves for three fullerites under hydrostatic compression: C$_{60}$ with simple cubic (SC) lattice, C$_{60}$ with face centered cubic (FCC) lattice and C$_{48}$ with SC lattice.

with SC packing differs significantly from fullerite with FCC packing of fullerenes. For structures with SC packing, the pressures obtained at the same densities exceed the pressure for the FCC fullerite. FCC packing of fullerene molecules is characterized by much more free space than with SC packing, which allows fullerenes to move in the lattice until high densities are reached. Fullerenes in the structure with FCC packing retain their spherical shape much longer than in the structure with SC packing.

The elastic moduli of single-crystal C_{60} were determined based on measurements of ultrasound velocities and were found to be of the order of $c_{11} = 15$ GPa, $c_{12} = 9$ GPa, and $c_{11} = 6$ GPa.

3.2.4.2 *Graphene Aerogel*

The compressibility and elasticity of graphene aerogels (GA) are mainly determined by the intrinsic feature of the building blocks, and the structural design of the porous network. GAs demonstrate high flexibility and superelasticity, but can be mechanically fragile under certain conditions, which leads to the collapse of the pores, changes of the specific surface area and consequently to property degradation. The existing reviews [268] cover a specific morphology, describing the syntheses, properties and the application, lacking in-depth discussions on the deformation mechanisms and mechanical properties in comparison. The graphene flake size, density of GA, and surface functional groups have a great effect on the mechanical performance of the GAs, especially on the compressibility and elasticity. The rigidity of the GA increases with its density, due to the formation of new van der Waals bonds and covalent bonds between neighboring graphene flakes. This behavior is explained by the special mechanical response of the graphene network to the compression. For compression to high density $\rho > 3.0$ g/cm^{-3}, amorphization of the GA occurs.

Figure 3.20 presents the stress-strain curves during uniaxial compression for cellular (C), honeycomb (H), and lamellar (L) GAs. Initial structures are presented in Fig. 2.30. Three initial structures are presented as the projection to xz-plane. Despite the cell shapes being very similar, the deformation behaviour is very different, especially under tension.

As can be seen, the pressure-density curves under compression for H and C coincide until $\rho = 3.0$ g/cm^3. The structural transformations during compression are presented as the snapshots of the structure at different compression densities. For honeycomb and cellular GA, zero pressure is observed up to $\rho = 2.2$ g/cm^3 which is the result of the continuous collapse of the hexagons (for H) or the opening of the hexagons (for C). During

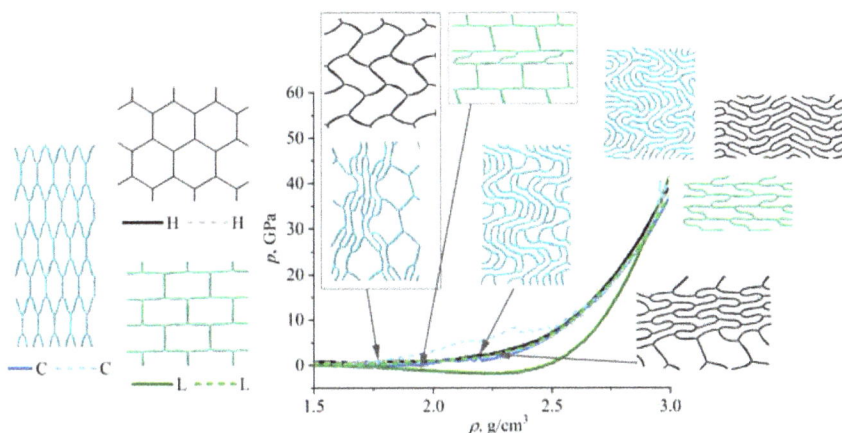

Fig. 3.20. (Left) Pressure-density curves during uniaxial compression: cellular (C), honeycomb (H), and lamellar (L) GAs. (Right) The snapshots of part of the structure as the projection to the xz-plane: initial state and at different densities.

compression, more and more hexagons collapse and rotate. For lamellar GA, the stress is even below zero, because during compression rectangles are collapsed line by line until $\rho = 2.7$ g/cm^3 (example of the first collapse is presented). The strain increase starts at $\rho = 2.2$ g/cm^3 when almost all the cells are collapsed. Strain increase is explained by the formation of van der Waals bonds between the flakes and the increase of their rigidity due to their folding. At $\rho = 3.0$ g/cm^3 all graphene walls are 3.4 Å, far from each other. Further strain increase leads to amorphization.

In comparison with the previously considered GA composed of graphene flakes, the slow stress increase can be explained by the pore removal during the first compression steps, and mutual alignment of the flakes in the structure and formation of folds and wrinkles. There is almost no covalent bonding in the structure; however, the number of covalent bonds across the edges of neighboring graphene flakes increases during compression.

Figure 3.21 presents the stress-strain curves for GAs under tension along the y and z axes together with the snapshots of the structure. UTS of H, L and C aerogels depends significantly on the tensile direction. Table 3.14 presents σ_{UTS}, fracture strain ε_F and Young's modulus.

As can be seen, the curves for H and L under tension along the y-axis and for H and C under tension along the z-axis can be divided into four regions: zero stress; linear elastic regime; inelastic deformation and pre-critical stress increase. These deformation stages are represented by blue regions in the

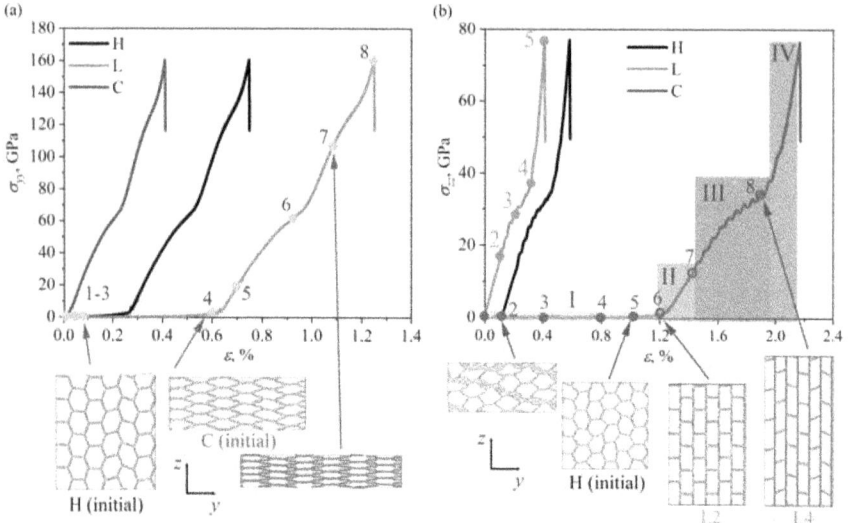

Fig. 3.21. Stress-strain curves under uniaxial tension along (a) y-axis and (b) z-axis for three GAs. Snapshots of the structure as the projection on yz-plane.

Table 3.14. Young's modulus E, ultimate tensile strength strength σ_{UTS}, fracture strain ε_F for GAs under tension.

Tensile direction	Structure	σ_{UTS}, GPa	ε_F	E, GPa
	H	160.25	67.38	308.19
y-axis	L	160.62	1.24	243.21
	C	160.40	0.40	361.15
	H	77.04	0.58	131.74
z-axis	L	77.11	0.41	145.62
	C	76.51	2.16	64.14

example of cellular GA under tension along z-axis. For cellular GA under tension along y-axis and lamellar GA under tension along z-axis, there is no region with zero stress level. All these differences and similarities are explained by the intrinsic structure of GA.

For deformation along y-axis, all the structural changes can be explained based on the snapshots for lamellar GA. At first, rectangular cells start to transform to the hexagonal (point 3, H initial). Oscillation of the hexagons took place up to $\varepsilon = 0.5$: some are rotated, some are opened, with continuous changing of their shape. Finally, all the cells transform to the hexagons and start to collapse during further tension. From point 3 we can describe the deformation behavior of the honeycomb GA (black curve in Fig. 3.21a) with the same snapshots of the structure. From point 4 we can describe the deformation behavior of cellular GA (dark-gray curve in Fig. 3.21a). At $\varepsilon = 0.6$, the hexagons cannot be collapsed any further and the graphene nanoribbons aligned with the tensile direction are stressed.

For tension along the z-axis the deformation behavior of H, L and C aerogels can be described by the same snapshots for cellular GA. The highest elasticity is found for cellular GA because during the first stages up to $\varepsilon = 1.2$, different structural changes took place: opening and rotation of the cells, oscillations of the graphene walls, transformation to hexagonal GA (point 5) and further transformation to lamellar GA (C curve, point 6; L curve, point 2). The most stressed walls are aligned with the tensile direction, while other graphene walls oscillate.

Figure 3.22 presents the overview of the fracture strain and strength of different aerogels for comparison. Five diamond nanomeshes [251], silica aerogel [199], GAs with different densities from [200], and continuous curved GAs [147] are compared by their UTS and strain. GAs studied in [200] are even weaker (pink stars) than those studied in the present work (red circles),

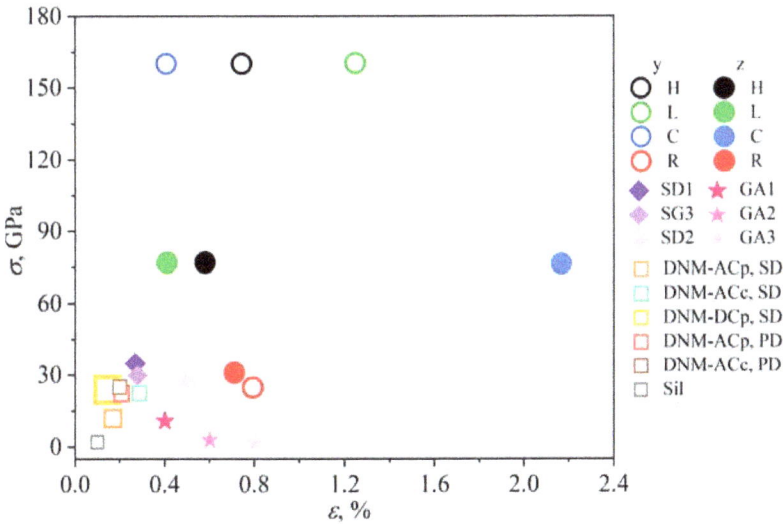

Fig. 3.22. Ultimate strength and fracture strain of GA configurations in comparison with other results (circles). Tensile strain and strength for SD1, SG3, SD2 are taken from [147], for GA1-GA3 from [200], for Sil from [199], for DNM from [251]. Reprinted with permission from [230].

because of the two-times-lower density of GA. In contrast, very similar discontinuous GAs failed by the fracture strain, because they have a more rigid structure with pre-defined interconnections between graphene flakes. It reveals that random distribution of graphene flakes in random GA results in better strength and fracture strain. Moreover, the higher the density of such GA, the higher the UTS under tension, but the lower the compressibility.

Various GAs composed of diamond nanothreads (square signs) are much weaker than pure graphene structures. Even changes in the structural morphology cannot result in strain increase. At the same time, honeycomb, cellular and lamellar GAs have a lot in common: high compressibility, high elasticity, and ability to transform from one structure to another. UTS and fracture strain, as well as the Young's modulus, are highly dependent on the tensile direction. The highest strength and fracture strain demonstrate the lamellar GA under tension along the direction perpendicular to the alignment of the graphene walls: this allows the transformation from lamellar to honeycomb and further to cellular structure. This also provides an opportunity to tune the structural morphology and properties of the GA.

3.2.5 *Diamond-Like Phases*

Figure 3.23 presents stress–strain curves under tension for all tubulanes, fulleranes, and graphene-based DLPs. Considered DLPs are presented in Figs. 2.33–2.35. The deformation of all tubulanes is very similar, with close limit of the first phase transformation. All the DLPs can be divided into three groups according to their deformation behaviour: (i) TA1, TA5, LA6, and CA3; (ii) TA3, TA6, TB, LA3, CA2, and CA8; (iii) CB, CA9, CA7, and TA8. For the third group, a sharp increase in stress at the last deformation stage is explained by a fast increase in all lattice parameters, while valence angles almost cannot be changed further. There is no clear phase transformation at this stage, just bond elongation. If compared with the values of Young's modulus, it can be seen that phases TA1, TA3, and TA5 possess a lower elastic modulus than TA7, TA8, and TB as well as lower critical strain and stress.

Tensile deformation is attributed to two deformation mechanisms – elongation and rotation of bonds. A strong correlation between the pressure–strain curve and changes in bond length and values of valence angles is observed. Also, there is a correlation with the stress distribution during tension: if the main contribution is because of the elongation of bonds, normal stress components are dominant; if the main contribution

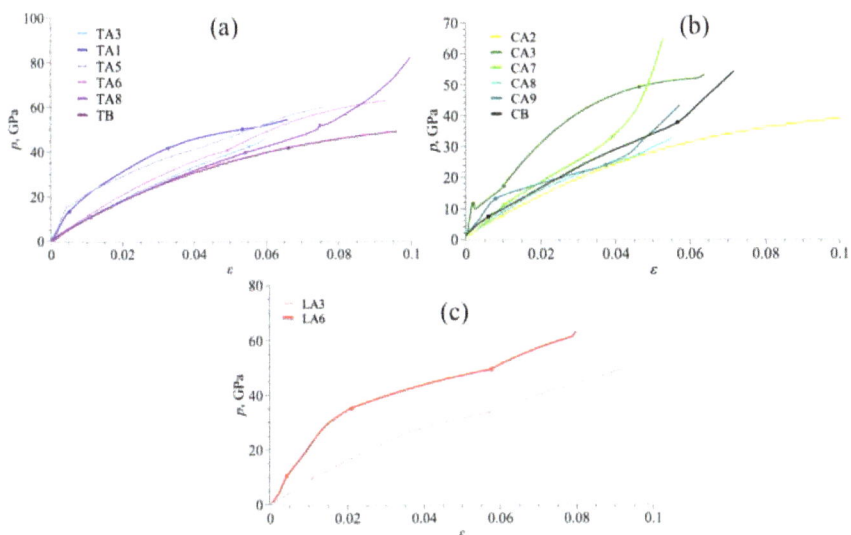

Fig. 3.23. Stress–strain curves under tension for (a) tubulanes, (b) fulleranes, and (c) DLPs based on graphene. Reprinted with permission from [12].

is because of changing valence angles, shear stress components are also meaningful. For structures of group I, the curve slope is much lower than for other DLPs. During the elastic region, for all DLPs, deformation is defined by changes in the lattice constants. Commonly, if the bond is aligned with the tensile direction, it decreases the strength of the DLP. On the contrary, if bonds are oriented at some angle to the tensile direction, deformation is defined by changes in valence angles which increase the final strength. In region II, a main contribution is made by a change in the lattice parameters (bond length increase for 1–4%). A slight change in the valence angles is found (about 1–2%). At stage II, changes in the lengths of valence bonds is the main deformation mechanism. Further, the main mechanism is the changes in valence bonds.

Figure 3.24 presents the pressure-strain curves for DLPs under compression: (a) for stable fulleranes, (b) for stable graphene-based DLPs, (c,d) for tubulanes. For all DLPs, compression occurred to densities of 3.13 g/cm^3, which is close to the density of diamond (3.47–3.55 g/cm^3).

As well as diamond, fulleranes remain stable up to high pressures with a maximum value of 120 GPa. Interestingly, the CA3 phase withstands the highest pressure level at the lowest critical stress. Almost all the curves

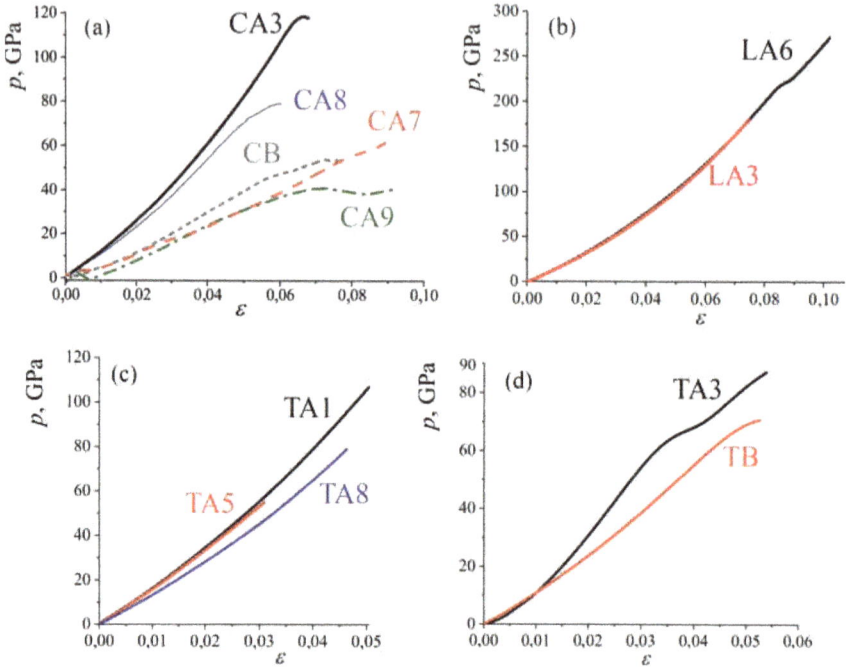

Fig. 3.24.   Pressure-strain curves for DLPs under compression: (a) for stable fulleranes, (b) for stable graphene-based DLPs, (c,d) for tubulanes.

have a small plateau, which indicates a transition to another structural state – amorphous. Two characteristic types of curves were found: with a small (CA3 and CA8) and a large (CA7, CA9 and CB) plateau. For the CA3 and CA8 phases, the amorphous state is observed at high pressures and densities close to the density of diamond. For the CA7, CA9 and CB phases, the transition to the amorphous state occurs already at low densities and a pressure of about 40-50 GPa.

From the analysis of stress components under hydrostatic compression it was found that the CA3, CA8, and CA9 phases deform differently than the CA7 and CB phases. The deformation for CA3 up to $\varepsilon = 0.06$ is realized mainly due to a decrease in bond lengths, while for higher deformations the shear stress components, initially close to 0.01 GPa, increase to 10 GPa at 0.073. For the CA7 and CB phases, the shear stress components were 12 GPa during the entire simulation. Thus, for the CA3, CA8, and CA9 phases, the deformation occurred due to a decrease in bond lengths, and for the CA7 and CB phases, the deformation occurred due to a decrease in bond lengths and a change in covalent angles.

Among tubulanes, similar deformation behavior can be distinguished for (i) TA1, TA5, TA8 and (ii) TA3, TB. Tubulanes, like fulleranes, remain stable up to high pressures with a maximum value of 115 GPa. The plateau on the curve, characteristic of fulleranes, is not observed, which indicates that tubulanes of the first group retain crystalline order up to high degrees of compression. The maximum value of deformation of 0.05 at $p = 115$ GPa (at 1 K) was found for TA1. For these tubulanes, no amortization was observed.

For graphene-based DLPs, no difference in the course of the curves is observed. The deformation of the LA3 phase is uniform and reaches a critical value at $\varepsilon = 0.075$, while for the LA6 phase in the deformation range of 0.085–0.09, a change in the deformation mechanism occurs and a small plateau appears on the curve. For both phases, compression was carried out to densities of 4.0 g/cm^3, which is close to the density of diamond at high pressure. Graphene-based DLPs remain stable even at high pressures with a maximum value of 180 GPa (for LA3) and 250 GPa (for LA6). For both phases the deformation mechanism is partly the reduction of valence bonds and partly the change of two covalent angles in the rhombic structural elements. A detailed analysis of the LA3 phase shows that the "cube" structural elements are transformed into rhombuses during relaxation, but at a high level of deformation the reverse transformation occurs.

3.2.6 *Composites*

3.2.6.1 *Fabrication Temperature*

Figure 3.25 presents an overview of the analysis of the effect of fabrication temperature on the process of composite fabrication. The temperature range from 0 to 5000 K is shown with some important points. Snapshots of the composites obtained by hydrostatic compression at 0 K are presented for small and large Ni nanoparticles inside the graphene flakes. Stress-strain curves for tension of the composites obtained at different temperatures from 1000 to 2000 K are also presented for the small size of the Ni nanoparticles.

From the comparison of the stress-strain curves for different sizes of Ni nanoparticles, it is evident that the structure is normally deformed to $\varepsilon = 0.3$. It can be seen that simple compression of the structure to high densities and strain does not lead to the formation of a composite: no chemical bonds appeared between the graphene flakes, which explains the

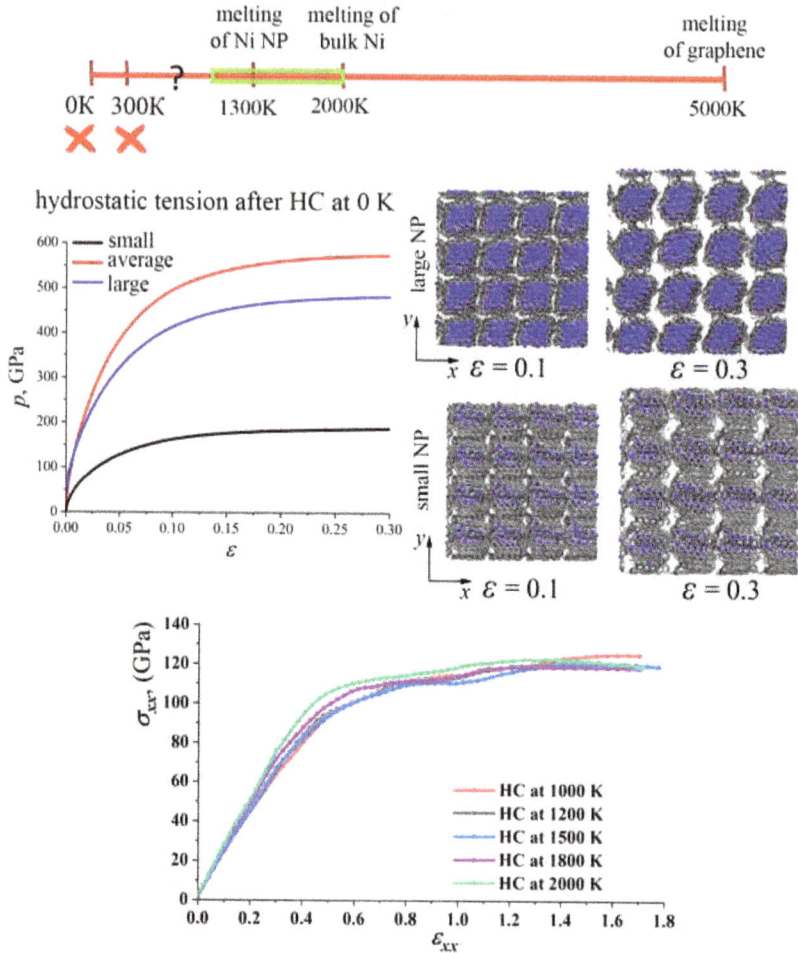

Fig. 3.25. (Top) Melting points presented along a temperature line. (Middle) The pressure-strain curves and snapshots of the composites obtained by hydrostatic compression at 0 K for small and large Ni nanoparticles inside the graphene flakes. (Bottom) The stress-strain curves for tension of the composites obtained at different temperatures from 1000 to 2000 K for small Ni nanoparticles. NP: nanoparticle; HC: hydrostatic compression.

plateau on the curve. At zero temperature the structural elements remain separated and, consequently, under tension pores appear very quickly and the material is transformed into the same system of flakes filled with particles as in the initial state.

Further tension results in the appearance of pores, because no chemical bonds between the atoms of neighboring graphene flakes appeared. Note that the larger the size of the nanoparticle, the greater the number and size of the pores in the structure. Thus, at 0 K it is impossible to obtain uniform composite structure. Hydrostatic compression at both 0 K and 300 K is ineffective for obtaining graphene/Ni composites. This is shown on the temperature line by red crosses.

To activate chemical bonds between individual elements of the composite, it is necessary to increase the deformation treatment temperature. The melting point of graphene is very high, about 5000 K. Moreover, the melting point of bulk metal and metal nanoparticles differs. The larger the nanoparticle size, the higher the melting temperature. Consequently, during deformation-thermal treatment at elevated temperatures, Ni nanoparticles will be in a pre-melting state at a temperature of 1000 K or in a molten state at higher temperatures from 1360 to 2000 K. The understanding of how temperature of the hydrostatic compression affects the composite fabrication is of high importnace. Figure 3.25 presents the stress-strain curves for tension of the composites obtained at different temperatures from 1000 to 2000 K for small Ni nanoparticles. As can be seen, there is no need to increase the temperature above 1000 K, since all the curves are similar. It is found that the Young's modulus of composites does not depend much on temperature. In the region of $0 < \varepsilon < 0.15$, the relationship between stress and strain is linear for all obtained composites. The calculated Young's modulus (averaged over five temperatures) for the graphene/Ni composite is 249 ± 7.

3.2.6.2 *Effect of the Nanoparticle Size*

The size of the metal nanoparticles also has a significant effect on the process of composite formation and its resulting strength. Figure 3.26 presents the stress-strain curves under tension for graphene network filled with Ni nanoparticles of different size. As can be seen, stress-strain curves for composites with small and average nanoparticles coincide up to $\varepsilon = 0.3$, which means that the strength of these composites is close, while the graphene/Ni composite with larger nanoparticles is significantly weaker. However, after the linear deformation mode, graphene network with average nanoparticles shows a decrease in maximum stress.

From Figure 3.26 it is evident that initially, the distribution of Ni atoms is much more uniform for the composite with small nanoparticles. Small nanoparticles are easily divided into individual atoms, which spread inside

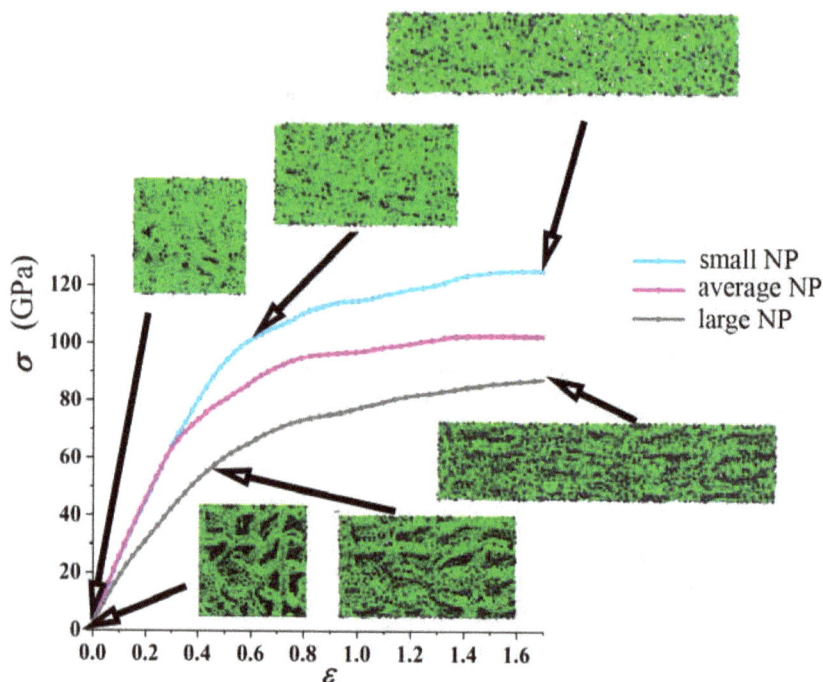

Fig. 3.26. Stress-strain curves under tension for three graphene/Ni composites with different sizes of Ni nanoparticles inside the pores. Snapshots of the composite under tension: green atoms for carbon, black atoms for Ni.

the graphene flakes. Thus, the composite with small nanoparticles is a more homogeneous structure.

The important factor is the ratio of the surface area of the graphene flake to the surface area of the nanoparticle. When the ratio of the free surface of graphene and nanoparticle is close to 1, all the free sites are occupied by metal atoms, which prevents the formation of a continuous graphene network, even for graphene/Ni composite. At the same time, Al and Cu coagulate into the large nanoparticles during hydrostatic compression, which weakens the composite. Moreover, under normal conditions, Al repels graphene and can come out onto the surface of the graphene network if it is allowed to do so. Thus, when graphene flakes are assembled into a network by deformation-temperature treatment, the pathways for Al to migrate out are closed, and it becomes possible to fabricate a graphene/Al composite. However, this composite is the weakest and most unstable.

3.2.6.3 *Different Metal Fillers*

The effect of metal type is of high importance for successful composite fabrication. There are two types of interaction between graphene and metal: weak and strong. Here, three different metals are considered – Ni which is strongly bonded to graphene, Cu which is more or less neutral, and Al which is weakly bonded to graphene. All the composites were obtained by hydrostatic compression at 1000 K for graphene/Ni composite, at 600 K for graphene/Cu composite, and at 300 K for graphene/Al composite. Figure 3.27 shows the stress-strain curves during uniaxial tension at 300 K for all the structures under consideration. Stress-strain curves for the three composites are shown in comparison with the stress-strain curves for graphene network obtained from the corresponding composite. Snapshots of the composites are presented on the left, while final structures at fracture moment are presented on the right. Graphene network without metal was considered to estimate the impact of the graphene network to the composite strength.

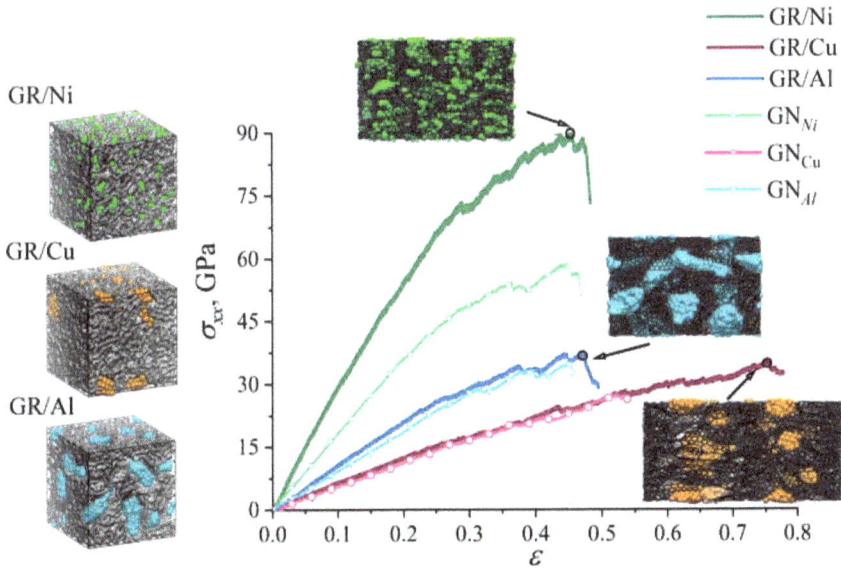

Fig. 3.27. Stress-strain curves during uniaxial tension at 300 K for three composites are shown by solid lines. For comparison, stress-strain curves for graphene network, obtained from the corresponding composite, are shown by solid lines with dots. Snapshots of the initial (left) and final (right) composite structures are also shown. GR: graphene; GN: graphene network.

Three different graphene networks were obtained for three different composites. The network obtained from graphene/Ni was more homogeneous, with small pores filled with planar metal nanoparticles. The metal nanoparticles even look like separate atoms, but this is just a problem of visualization: Ni nanoparticles are small, but still keep their unity and fill the pores of the graphene network. For graphene/Cu and graphene/Al composites, graphene networks with large pores were formed due to the coagulation of metal nanoparticles. For graphene/Al composite, coagulation of Al nanoparticles was even more pronounced.

Let us first discuss the composite behavior from the stress-strain curves. As can be seen from Figure 3.27 (solid lines), graphene/Ni composite has the highest UTS and the lowest fracture strain or lowest plasticity. Composites graphene/Cu and graphene/Al have similar UTS but different fracture strain (about two times higher for graphene/Cu).

Table 3.15 shows the UTS and fracture strain for all the composites and graphene networks.

For graphene/Ni composite, the homogeneous graphene network was obtained during high-temperature hydrostatic compression due to the strong interaction between graphene and Ni. The homogeneous distribution of metal atoms can be seen from the initial structure. Pores appeared in the structure only at very high strain (about $\varepsilon \approx 0.42$). A continuous rearrangement of the graphene network took place under tension (covalent bonds were destroyed and new covalent bonds were formed); however, in such a dense graphene network with a large number of covalent bonds, stress was accumulated, which did not allow high ductility.

Comparison of the stress-strain curves for graphene/Ni and graphene network$_{Ni}$ showed a much lower UTS for the latter (Figure 3.27). When Ni

Table 3.15. Ultimate tensile strength σ_{UTS}, fracture strain ε_F and Young's modulus E of the structures under consideration.

Structure	σ_{UTS}, GPa	ε_F	E, GPa
GR/Ni	89.5	0.45	296.9
GR/Cu	35.1	0.75	65.9
GR/Al	36.8	0.47	109.1
GN	52.0	0.63	140.7
GN$_{Ni}$	58.1	0.45	179.2
GN$_{Cu}$	27.6	0.52	55.3
GN$_{Al}$	34.6	0.45	91.4

atoms were removed from the structure, which was followed by relaxation, graphene network$_{Ni}$ became closer to the graphene network of amorphous carbon. Pores, where nanoparticles were deposited, healed during relaxation with the formation of new covalent bonds. As a result, the brittleness of the composite increased. Note that Ni nanoparticles can also anchor the neighboring graphene flakes due to the high binding energy, which also increases the composite strength.

The Young's modulus E can be defined from the linear regime of the stress-strain curves under tension (Fig. 3.27). Graphene/Ni composite has the highest Young's modulus equal to 296.9 GPa. For graphene/Al and graphene/Cu composites the Young's moduli are $E = 109.1$ GPa and $E = 65.9$ GPa, respectively. The comparison of Young's moduli shows that the graphene/Ni composite has higher stiffness than other composites.

No such difference was found for graphene/Cu and graphene/Al composites and their graphene networks. After the removal of metal nanoparticles and relaxation of the graphene network, the structure was slightly rearranged, but all the pores were preserved. For the graphene/Cu composite, the stress-strain curves for the composite itself and the pure graphene network$_{Cu}$ almost coincide up to $\varepsilon = 0.5$ (Figure 3.27). After the removal of Cu nanoparticles from the graphene network, large pores appeared, resulting in a decrease of the tensile strength. Moreover, Cu nanoparticles act as lubricants for better plasticity of the graphene/Cu composite.

An important difference should be noted between graphene/Cu and graphene/Al composites: the distribution of the metal nanoparticles during tension. For graphene/Cu, Cu nanoparticles look separate, while for graphene/Al composite, the Al nanoparticles are connected to a metal network. Consequently, for the graphene/Cu composite, pores appeared in the structure at a much earlier deformation stage (about $\varepsilon = 0.14$). Stress-strain curve before $\varepsilon = 0.42$ is almost linear with slight oscillations. Pores appear at the interface between Cu and the graphene network, but this does not lead to composite fracture. Due to the rearrangement of carbon bonds, the composite remains stable, and the fracture occurred just when this pore became too large. The graphene network during tension looks like separate graphene flakes interconnected at the edges, which increases the composite ductility.

For the graphene/Al composite, the first pores appear at $\varepsilon = 0.36$. As with graphene/Cu, pores appear at the interface between Al and the graphene network; however, it does not lead to fracture of the structure due to rearrangement of carbon bonds. The graphene network under tension looks like separate graphene flakes connected at the edges. Al nanoparticles tend to coagulate easily, forming larger and larger clusters. This process prevents the formation of new carbon bonds. In contrast, Cu nanoparticles interact less with each other and more strongly with the graphene network. As a result, both the graphene network and the copper nanoparticles are deformed during tension of the graphene/Cu composite. The latter contribute significantly to the ductility of the composite.

Although graphene network was obtained during hydrostatic compression of graphene/Al and graphene/Cu composites even at low temperature, which is an advantage, this graphene network could not withstand fracture at the graphene/metal interface under tensile strain. The fracture occurs exactly at the interface between metal and graphene. The low tensile strength compared to the graphene/Ni composite can also be explained by the formation of sufficiently large metal nanoparticles during hydrostatic compression. As a result, a large surface area was obtained between the metal nanoparticle and the graphene network. For graphene/Ni, the metal is very homogeneously distributed inside the network compared to graphene/Cu and graphene/Al. Note that the graphene network is closely intertwined with the Ni nanoparticles which is induced by the strong Ni–C interactions.

3.2.6.4 *Functionalization of Graphene Network*

To date, numerous experimental and theoretical studies have been conducted to synthesize and analyze the properties of various graphene/metal composites; however, there is still a long way before the understanding of their properties and morphology control. Numerous problems arise during the fabrication process: low interface strength between graphene and metal, inhomogeneous distribution of the graphene reinforcement in the matrix, appearance of defects or other chemical additions during synthesis.

The wettability of graphene with metal and their adhesion is one of the important problems which should be solved to enhance interfacial bonding strength in such composites. It has been shown that graphene/metal interaction can depend on very different factors: defects, graphene curvature, presence of other atoms, etc. Graphene functionalization is one of the effective ways to enhance graphene adhesion to metal. One of the possibilities to increase the strength of the graphene/metal interface is graphene decoration by metal atoms; for example, nickel (Ni) strongly interacts with graphene flakes, can be easily anchored to graphene and affect the bonding between graphene and metal.

Figure 3.28 presents the pressure-density curves obtained during the compression of the composite precursor: graphene network filled with metal (Cu and Al) nanoparticles. These metals were chosen because of low adhesion to graphene. To enchance the interface bonding, graphene flakes were decorated with Ni nanoclusters. Two curves are compared for Al/graphene and Al/graphene decorated with Ni, as well as two curves for Cu/graphene with and without Ni. Temperature of the hydrostatic compression is 300 K. Pressure in the system is zero while the structural units are far away from each other, and only start to interact at $\rho = 1\ \mathrm{g/cm^{-3}}$. The rigidity of the composite increases with its density, due to the formation of new van der Waals and covalent bonds between neighbouring graphene flakes and new covalent bonds at the edges (more details on structure transformation are given below).

Snapshots of the structure during compression are also presented in Fig. 3.28. As can be seen, Cu nanoparticles are covered with graphene in both cases; however, in the presence of Ni, a denser structure was obtained. In the structure without Ni, Cu nanoparticles can coagulate when the

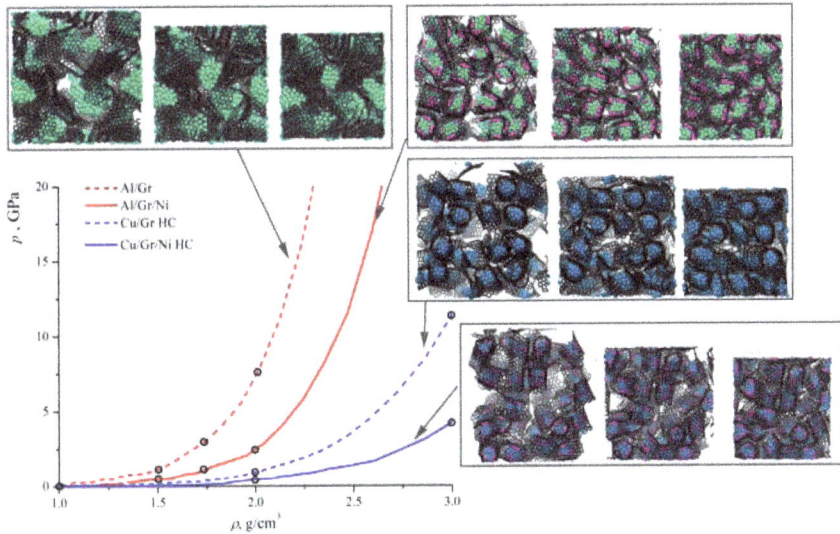

Fig. 3.28. Stress-strain curves under hydrostatic compression at 300 K. Snapshots of the structure during compression: black atoms for graphene, green for Al, blue for Cu, and purple for Ni. Gr: graphene; HC: hydrostatic compression.

flake edges are opened, while for Cu/graphene/Ni, coagulation of metal nanoparticles is impossible.

Rapid increase of the pressure for Al/graphene (Al/graphene/Ni) is explained by the strong repulsion of Al and graphene: even at high compression level, Al nanoparticles are placed at about 5-7 Å from graphene flakes. For Al/graphene, even the initial structures of the composite precursors obtained during initial relaxation are different. For Al/graphene, a graphene network with large surface area, filled with large Al nanoparticles, was obtained with non-homogeneous distribution of Al nanoparticles. The presence of Ni decoration results in a considerable anchoring of Al nanoparticles to graphene. In the structure with Ni, graphene flakes are curved, strongly anchored with metal nanoparticles, monotonously distributed inside the graphene network. Ni decoration increases the interfacial strength between Al and graphene, but can prevent the formation of a strong covalent network.

Analysis of the Ni-decorated composites under tension demonstrated that the strength of the Cu graphene composite with Ni decoration is much lower than that without Ni. The same was found for the Al/graphene system. For Cu/graphene composite the Young's modulus is equal to 18 GPa. Young's modulus for Cu/graphene/Ni, Al/graphene and Al/graphene/Ni are close and equal to 12 GPa. Interestingly, the presence of Ni increases the composite ductility, but decreases its strength. The structural analysis showed that without Ni the obtained graphene network has better interconnection, with a larger number of atoms with sp^3 hybridization. The reason to add Ni to the structure was to increase the metal/graphene interface strength for weakly interacting Al and almost neutral Cu. Analysis of the structure under compression showed that Ni definitely increases the interaction between Al/graphene and Cu/graphene, but at the same time prevents the formation of a covalent graphene network. The structural elements interact mainly with van der Waals bonds, which results in a weakening of the composite.

3.3   Link Between Structure and Properties

3.3.1   *Elastic Strain Engeneering*

Nanostructured materials (such as thin films, nanofibers, nanoparticles, and bulk nanomaterials) can exhibit high strengths up to the theoretical strength without stress relaxation through inelastic deformation or rupture.

Large elastic deformations of up to 10% can be achieved for nanomaterials, unlike traditional materials. This feature opens up new possibilities for controlling the physical properties of the material, such as electronic, optical, magnetic, phononic and catalytic, while only controlling the components of the applied strain tensor. Such a simple and effective method for changing and controlling the physical properties of graphene and other nanosized carbon structures is called **elastic strain engineering**.

High-pressure physics has shown that unique physical properties can be achieved by simply varying stresses. It is well known that all properties of a crystalline material strongly depend on the lattice parameter and the shape of the primitive cell, which can be changed by applying stresses. In recent decades, it has been discovered that elastic strains, such as shear or uniform tension, can be successfully used to control the properties of nanomaterials. Previously, the main difficulty was the possibility of compensating for shear and tension by plastic deformation or fracture of the material. Traditional materials cannot withstand elastic strains above 0.2–0.3%, since at higher values, inelastic relaxation occurs. However, the development of nanotechnology in recent decades has made it possible to obtain qualitatively new nanostructures (which are also called superstrong), capable of withstanding tensile stresses of over 1% for a fairly long time. In addition to elastic deformation, properties can be controlled by applying plastic deformation.

Elastic strain engineering represents a qualitatively new level of control over the properties of a material. When researchers talk about the use of deformation to change properties, they usually mean changing the mechanical properties, rather than the physical or chemical ones. Elastic deformation technology, in turn, leads to a change in both properties. Since almost all physical and chemical properties of a material depend on its electronic structure, and the electronic structure changes greatly near the stability limit of the material, the properties near this limit should differ greatly from the properties in the unstressed state. As a theoretical concept, elastic strain engineering is not new, and has previously been considered by many scientists as a possible way to change properties; however, its application seemed impossible until new nanostructured materials appeared that showed significant changes in properties when even small deformations were applied.

For the application of elastic strain engineering, it is very important that the material can withstand sufficiently large elastic deformations, as well as at low defect concentration. Another important factor is the processing

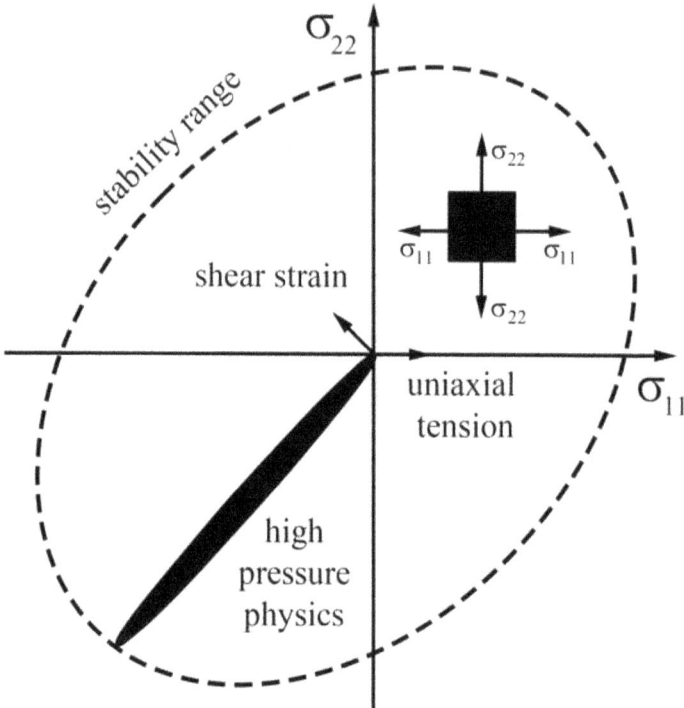

Fig. 3.29. Elastic strain engineering: schematic.

temperature. At low temperatures, the principle that the smaller the size, the stronger the material is justified for most materials, but at elevated temperatures, due to the activation of diffusion creep, this principle is not fulfilled. Thus, elastic strain engineering should be applied to materials with a high melting point, since the critical temperature of the transition from a material with high mechanical properties to a weak one strongly depends on its melting temperature. At the same time, devices based on such nanostructures are mainly used at room temperatures or below.

Currently, elastic strain engeneering is actively used to change the properties of carbon nanomaterials. For example, it was found that the thermal conductivity of graphene and CNTs decreases monotonically with increasing tensile stresses. The optical conductivity of graphene also depends significantly on deformation. The change in the electron spectrum during elastic in-plane deformation of graphene was considered, which showed that the formation of an energy gap in the electron spectrum of graphene occurs for a deformation of the order of 0.15 and has a threshold character for different directions of deformation. First-principles modeling also shows that uniaxial deformation up to 0.1 does not lead to the occurrence of a gap in the spectrum of phonon oscillations of graphene atoms.

3.3.1.1 *Orientation Dependence of the Sound Velocities of Graphene*

Figures 3.30-3.32 show the orientation dependence of the sound velocities of graphene under tension for loading along zigzag, armchair and mixed direction, respectively. The strain components are indicated in the captions.

These results reflect the symmetry of strained graphene. For hydrostatic loading, there is no dependence of the sound velocities on the direction; properties are isotropic. In Fig. 3.30 (tension along the armchair direction) and Fig. 3.31 (tension along the zigzag direction), graphene is orthotropic. The orientation dependence of the sound velocities for loading with nonzero shear components is shown in Fig. 3.32. Note that near the boundary of the stability region one of the sound velocities becomes zero.

3.3.1.2 *Gap in the Density of the Phonon States*

Figure 3.33 presents the stability region of flat graphene and the density of the phonon states (DOS) at different strain. As can be seen from the DOS, point A, no gaps exist for unstrained graphene, and a gap opens in graphene uniaxially loaded along the zigzag or armchair direction. At points B, C and F, DOS are presented for strains close to ripple formation and here the gaps

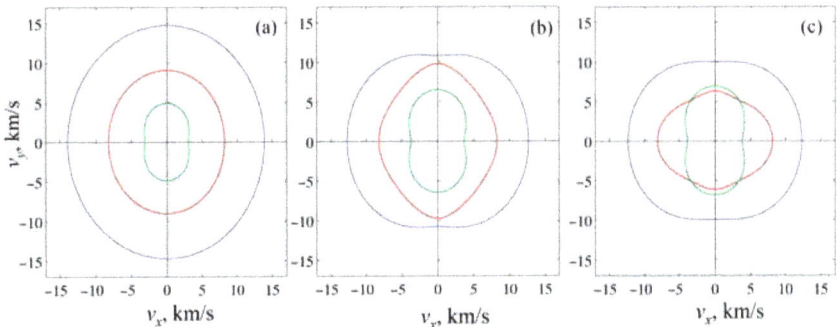

Fig. 3.30. Orientation dependence of the sound velocities of graphene under tension along armchair direction: (a) $\varepsilon_{yy} = 0.1$; (b) $\varepsilon_{yy} = 0.2$; (c) $\varepsilon_{yy} = 0.25$. Here $\varepsilon_{xx} = \varepsilon_{xy} = 0$.

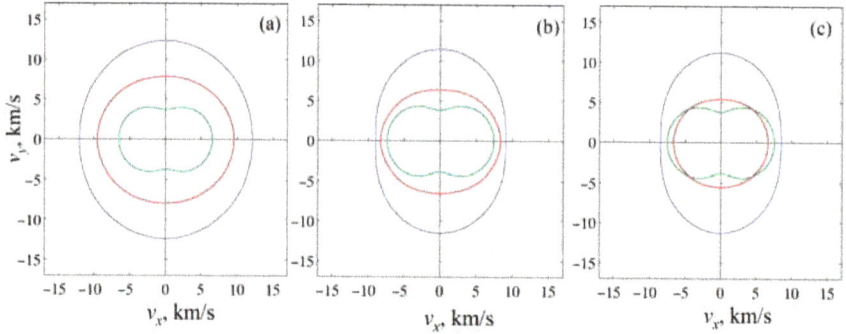

Fig. 3.31. Orientation dependence of the sound velocities of graphene under tension along zigzag direction: (a) $\varepsilon_{xx} = 0.2$; (b) $\varepsilon_{xx} = 0.29$; (c) $\varepsilon_{xx} = 0.32$. Here $\varepsilon_{xx} = \varepsilon_{xy} = 0$.

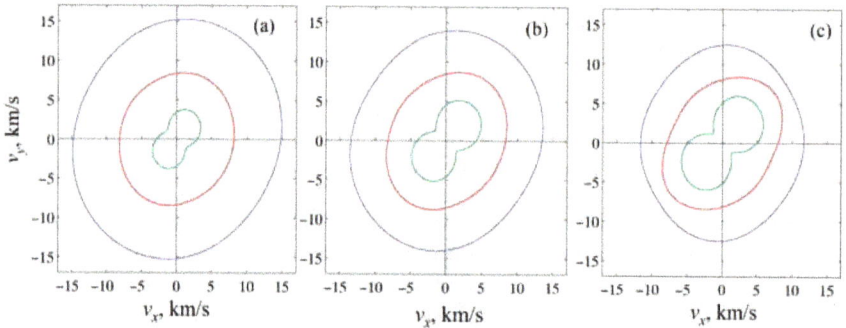

Fig. 3.32. Orientation dependence of the sound velocities of graphene under tension at: (a) $\varepsilon_{xy} = 0.1$; (b) $\varepsilon_{xy} = 0.2$; (c) $\varepsilon_{xy} = 0.25$. Here $\varepsilon_{xx} = 0.11\varepsilon_{xy}$, $\varepsilon_{yy} = 0.45\varepsilon_{xy}$.

are more prominent, especially for positive strain along zigzag direction is in combination with negative strain in the armchair direction. Narrower gaps are observed in D and E when positive strain along armchair direction is combined with negative or zero strain in zigzag direction. Thus, elastic strain can be effectively applied to open the gap. Existence of gaps in the phonon spectrum is important for the possibility to further use graphene in nanoelectronics.

3.3.2 Non-Elastic Deformation

Let us consider the effect of temperature and initial structure on the deformation behaviour of the 3D graphenes of different morphology.

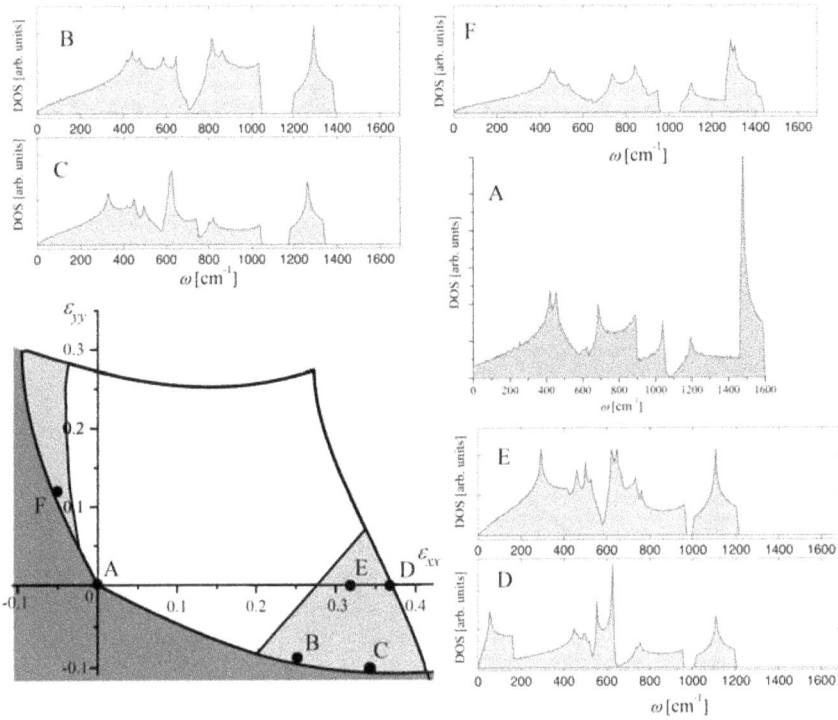

Fig. 3.33. Stability region of flat graphene. Dark-gray area – ripple formation, light-gray area – appearance of the band gap in DOS. DOS at different strain: A $\varepsilon_{xx} = \varepsilon_{yy} = 0$, B $\varepsilon_{xx} = 0.25$, $\varepsilon_{yy} = -0.1$, C $\varepsilon_{xx} = 0.35$, $\varepsilon_{yy} = -0.1$, D $\varepsilon_{xx} = 0.368$, $\varepsilon_{yy} = 0$, E $\varepsilon_{xx} = 0.32$, $\varepsilon_{yy} = 0$, F $\varepsilon_{xx} = -0.05$, $\varepsilon_{yy} = 0.12$.

Four types of bulk carbon nanomaterials, named bulk graphene composed of graphene flakes (BGF), bulk graphene composed of CNTs (BCNT), bulk fullerite composed of fullerenes C_{240} (BF) and bulk graphene composed of a mixture of graphene flakes, CNTs and fullerenes (BMIX). These structures were considered earlier in this chapter. Hydrostatic ($\varepsilon_{xx} = \varepsilon_{yy} = \varepsilon_{zz} = \varepsilon$), biaxial ($\varepsilon_{xx} = \varepsilon_{yy} = \varepsilon$ and $\varepsilon_{zz} = 0$) and uniaxial ($\varepsilon_{xx} = \varepsilon$ and $\varepsilon_{xx} = \varepsilon_{zz} = 0$) compressions were applied to the computational cell.

Figure 3.34 presents the pressure (stress) as a function of density for the four structures, (a) BGF, (b) BCNT, (c) BF and (d) BMIX, at $T = 1500$ K. All the structures are compression-resistant just like paper balls as compressive stress makes them stiffer and harder. For BGF the effect of the loading scheme is the weakest, while for BF there is a significant

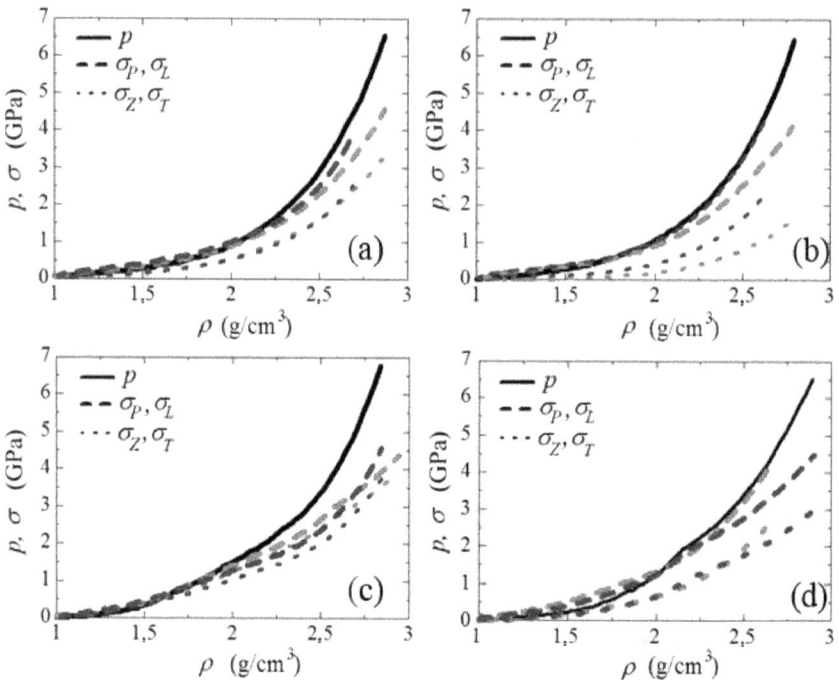

Fig. 3.34. Pressure (stress) as a function of density at $T = 1500$ K for the four materials (a) BGF, (b) BCNT, (c) BF and (d) BMIX for three loading schemes: hydrostatic compression (solid lines), biaxial compression (gray dashed and dotted lines for σ_P and σ_Z, respectively) and uniaxial compression (black dashed and dotted lines for σ_L and σ_T, respectively). Reprinted with permission from [11].

difference between the curve for the hydrostatic and those for the other two types of compression. For BCNT, curves for hydrostatic and uniaxial compressions coincide, but differ from that for biaxial compression. Again, graphene flakes can easily bend, fullerenes sustain large strain and CNTs can be easily collapsed, but hardly bent along the direction of the CNT axis.

A temperature range of 300–3000 K for the three loading schemes was considered (see Fig. 3.35). Under hydrostatic compression, BF shows the highest strength, while BGF shows the lowest. At higher temperatures the calculated stresses are slightly higher. For $\rho < 1.5$ g/cm^3 at high temperatures the strength of BF is the lowest, while for $\rho > 1.5$ g/cm^3

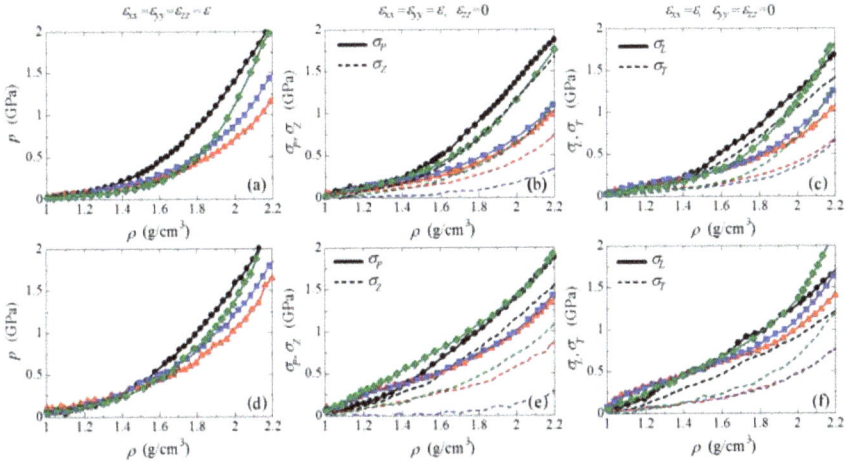

Fig. 3.35. Pressure (stress) as a function of density for BGF (red triangles), BCNT (blue squares), BF (black dots) and BMIX (green rhombuses) at two temperatures, $T = 300$ K (a–c) and $T = 3000$ K (d–f). Three loading schemes: (a, d) hydrostatic, (b, e) biaxial and (c, f) uniaxial compressions. Reprinted with permission from [11].

BF shows the highest strength. At high temperatures the graphene flakes and CNTs rotate during compression and can be deformed more easily. The strengthening of BGF and BCNT for $\rho > 1.5$ g/cm^3 occurred.

Fullerenes at high temperatures can easily move in the simulation box without any interaction with each other because the porosity of BF at $\rho = 1$ g/cm^3 is much higher than those for BGF and BCNT. At $\rho > 1.5$ g/cm^3 the fullerenes in BF start to interact which results in an increase in strength that then becomes higher than for BGF and BCNT. For all loading schemes at $\rho = 2$ g/cm^3 BF had the highest strength and BGF had the lowest strength.

The mixture of different structural elements BMIX showed an average behavior. For the biaxial and uniaxial compressions the strength increased with the temperature for BMIX and became even higher than the strength of BF. Interestingly, the addition of shear stress to the hydrostatic pressure considerably reduces the pressure needed to destroy the fullerene cages.

Generally applicable constitutive relationships describing the deformation of bulk carbon nanostructures, including crumpled graphene, can be

very helpful, because the properties of such materials are highly dependent
on the material processing history [11].

The pressure–density (stress–density) nonlinear curves can be fitted well
to the power law

$$\sigma = A\frac{\rho}{\rho_0}^{\alpha},$$

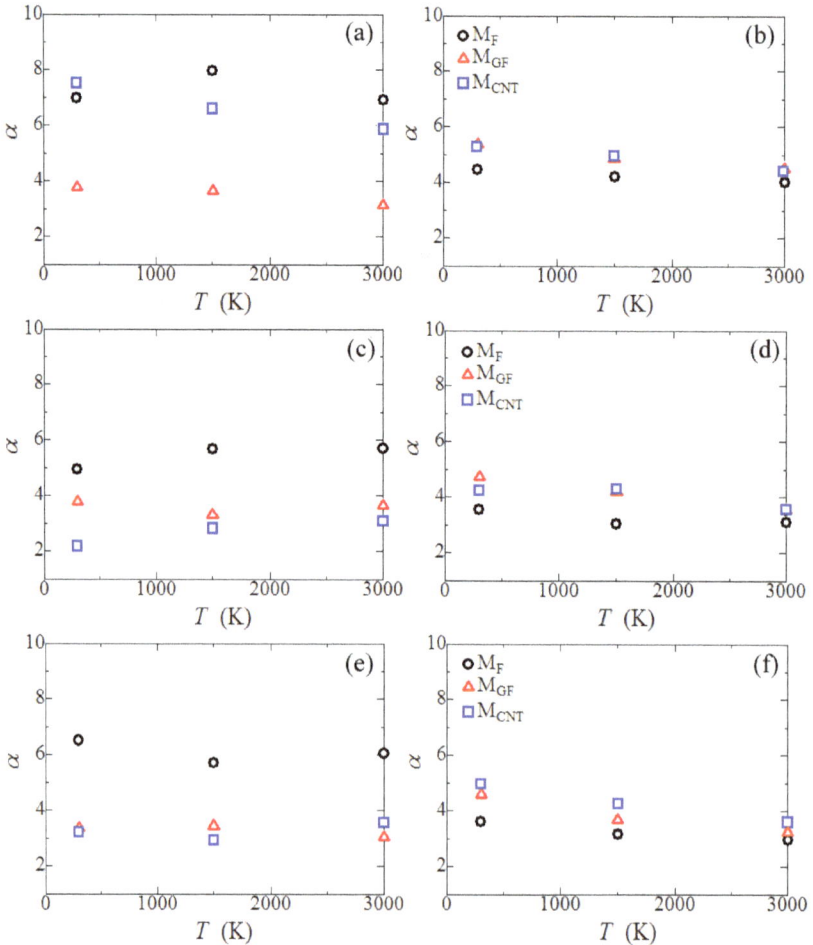

Fig. 3.36.   Coefficient α as a function of temperature for BGF, BCNT and BF at (a,c,e)
$1 < \rho < 1.5$ g/cm^3 and (b,d,f) $1.5 < \rho < 3.0$ g/cm^3 under (a,b) hydrostatic, (c, d)
biaxial and (e, f) uniaxial compressions. Reprinted with permission from [11].

for the hydrostatic, biaxial and uniaxial compressions. Here σ is measured in Pa, ρ in g/cm^3, A and α are parameters and $\rho_0 = 1$ g/cm^3 is the initial density. To achieve a better fit the constants A and α were assumed to be different for the two density ranges $1 < \rho < 1.5$ g/cm^3 and $1.5 < \rho < 3.0$ g/cm^3.

In Fig. 3.36, the coefficient α as a function of temperature at (a, c, e) $1 < \rho < 1.5$ g/cm^3 and (b, d, f) $1.5 < \rho < 3.0$ g/cm^3 is shown for BGF (red triangles), BCNT (blue squares) and BF (black dots). Hydrostatic (a, b), biaxial (c, d) and uniaxial (e, f) compressions are presented. BMIX was not considered in this case because of its complex behavior which can be described by the mechanical response of its building units. It can be seen that for the density range $1 < \rho < 1.5$ g/cm^3 the coefficients of the power functions are different for all materials, while for $1.5 < \rho < 3.0$ g/cm^3 they are almost the same, which means that at high densities the strain response is very similar. Moreover, for $1.5 < \rho < 3.0$ g/cm^3 the difference between the α values decreases with an increase of temperature. For $1.5 < \rho < 3.0$ g/cm^3 BF shows the highest rate of strengthening for all loading schemes, and BGF and BCNT show the lowest and almost equal rates of strengthening. For $1 < \rho < 1.5$ g/cm^3, under hydrostatic compression (Fig. 3.36a) α slightly decreases with an increase of temperature, while under biaxial compression it slightly increases. Under uniaxial compression α is almost the same for all the temperatures. The fitting accuracy of these constants is within 8%.

Chapter 4

Molecular Dynamics Study of the Mechanical Properties

One of the main problems in the investigation of the mechanical properties of graphene or other nanostructured materials is their small size, which do not allow us to effectively use experimental techniques. Several interesting methods were developed to analyze the mechanical properties of graphene, described in Chapter 2. However, most of the results were obtained by different simulation techniques like *ab initio* or molecular dynamics (MD) simulation. In this chapter, the methodology of the calculation and analysis of the mechanical properties of carbon nanostructures will be presented.

In MD simulation there are also a lot of different peculiarities which can considerably affect the results. Thus, it is very important to understand how to conduct MD simulation, how to choose the boundary conditions, simulation cell, interatomic potential, to name a few. Figure 4.1 shows some important issues, which should be taken into account for MD simulations.

4.1 Molecular Dynamics

MD simulation is based on the mathematical description of the interaction between atoms. The accuracy of predictions made based on the simulation results depends on the accuracy of this description and application to a specific problem. Classical methods of interaction are described using the potential function $U(\vec{r}_1, \vec{r}_2, \ldots, \vec{r}_N)$, which determines the potential energy of a system of N atoms as the function of their coordinates. The dynamics of atoms in the system is governed by Newton's equation of motion and the

ABOUT MD

How to apply deformation?

Covalent bonding
Van-der-Waals interaction
Rotation

$U = U_1 + U_2 + U_n$

$U_n = f(?)$

Interatomic potential

Boundary
conditions?

Strain rate?
Time step?
Cut-off?

- Should we apply strain or
stress?
- NVT or NPT?

Details

Scale problem: properties of
nanostructures considerably
dependent on its size

Comparison problem: where the
experimental results?

Experiment?

and even more problems...

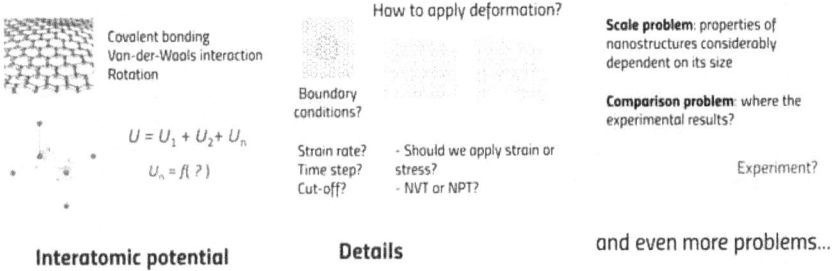

Fig. 4.1. MD for the simulation of carbon nanostructures.

forces acting on each atom are calculated as:

$$\vec{F}_i = -\frac{\partial U(\vec{r}_1, \vec{r}_2, \ldots, \vec{r}_N)}{\partial \vec{r}_i} \equiv -\nabla_i U(\vec{r}_1, \vec{r}_2, \ldots, \vec{r}_N),$$

$$\vec{a}_i = \frac{\vec{F}_i}{m_i},$$

where $\vec{F}_i$, $\vec{a}_i$ and m_i are the force, acceleration, and mass of atom i, respectively, and U the potential energy of the system.

The simplest form of description of interatomic interaction is pair potentials. Strictly speaking, these potentials do not have quantum mechanical justification. However, due to their simplicity, they are often used in modeling. At approximation of pair potentials, the energy of a system of particles is represented as the sum of potential energy interactions of all pairs of atoms:

$$U(\vec{r}_1, \vec{r}_2, \ldots, \vec{r}_N) = \frac{1}{2} \sum_{i=1}^{N} \sum_{j=1,(j\neq i)}^{N} \varphi(r_{i,j}),$$

where $r_{i,j} = |r_j - r_i|$ is the distance between a pair of atoms. For any of the considered structures it is necessary to choose a potential function that will properly describe the atomic interaction.

The main steps of the application of MD are (Fig. 4.2):

(i) Description of interatomic interaction;

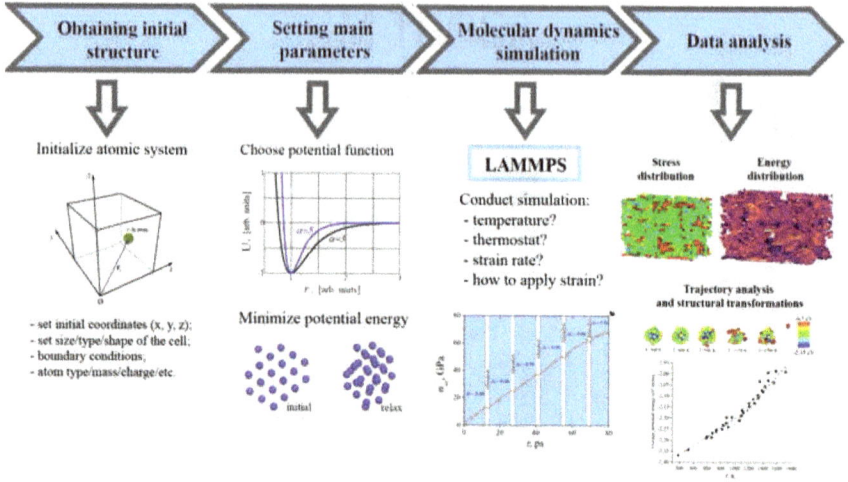

Fig. 4.2. MD simulation chart on how to conduct the simulation.

(ii) Initialization of the atomic system for simulation (to choose the simulation cell);

(iii) Carrying out MD calculations (simulation conditions or type of statistical ensemble, achievement of thermodynamic equilibrium, number of MD steps, etc.);

(iv) Analysis of simulation results (calculation of physical quantities, visualization of results, etc.).

There are a lot of works on the detailed description of the methodology of MD simulation. Here, just the practical examples would be shown for readers who are familiar with this method and can use the presented examples for their work. All the examples are presented for the LAMMPS simulation package. However, all the required interatomic potentials are described below for convenience.

4.1.1 *Simulation Tools*

One of the most common integration mechanisms is the velocity-Verlet algorithm, because it requires less computer memory. For better accuracy, velocities of the atoms are updated at half-integer time steps, and the new positions are defined as

$$\vec{v}_i\left(t + \frac{\delta t}{2}\right) = \vec{v}_i\left(t - \frac{\delta t}{2}\right) + \frac{\vec{F}_i(t)}{m_i}\delta t;$$

$$\vec{r}_i(t + \delta t) = \vec{r}_i(t) + \vec{v}_i\left(t + \frac{\delta t}{2}\right)\delta t,$$

where δt is the timestep. The velocity of the atom at time t will be defined as

$$\vec{v}_i(t) = \frac{1}{2}\left[\vec{v}_i\left(t - \frac{\delta t}{2}\right) + \vec{v}_i\left(t + \frac{\delta t}{2}\right)\right].$$

In this case, the atomic velocities and positions are not defined at the same time; kinetic and potential energies also cannot be calculated simultaneously. Thus, the velocity-Verlet algorithm is proposed:

$$\vec{v}_i\left(t + \frac{\delta t}{2}\right) = \vec{v}_i(t) + \frac{1}{2}\frac{\vec{F}_i(t)}{m_i}\delta t,$$

$$\vec{r}_i(t + \delta t) = \vec{r}_i(t) + \vec{v}_i\left(t + \frac{\delta t}{2}\right)\delta t,$$

$$\vec{v}_i(t + \delta t) = \vec{v}_i\left(t + \frac{\delta t}{2}\right) + \frac{1}{2}\frac{\vec{F}_i(t + \delta t)}{m_i}\delta t.$$

The error in the velocity-Verlet algorithm is of the same order as that in the Verlet scheme.

The first stage of an MD simulation is the building of the molecular system which will closely represent some experimental structure under investigation. It should be as realistic as possible. As a result of MD simulation, the trajectories of all N atoms in the system will be produced as a function of time. The final trajectories will considerably depend on the thermodynamic properties, correlation functions, and transport properties of the system under consideration. Thus, all the internal and external factors are highly important to obtain physically realistic results. Conditions such as the configuration of the system, velocities, the type of ensemble, the boundary condition, the numerical algorithm, etc., have to be properly chosen at the first step of the simulation.

The boundary conditions allow us to simulate different systems. For example, to reproduce the 3D bulk crystal, periodic boundary conditions are usually applied: the interactions between atoms can be across the boundary, and atoms can freely exit one side of the box and re-enter from the other side. In the case of the simulation of isolated systems like nanoparticles, atoms do not interact across the boundary and the nonperiodic boundary condition can be applied. In this case, a simulation cell with size much larger than the simulated object can be used.

The Nose-Hoover thermostats will be further used for all the presented examples to re-scale the velocities of atoms in MD simulations to control the temperature [102].

Most of the MD packages, such as LAMMPS [209], CHARMM [33], AMBER [38], NAMD [186] and GROMACS [1] can be successfully used for the simulation of very different phenomena for a wide range of materials: pure metal, carbon structures, composites, polymers, biological systems, to name a few.

Some of these MD packages even allow us to generate simple structures with a pre-defined crystal structure, like FCC, BCC or HCP atomic order for metals. More complex structures, described further, can be obtained also with some visualization tools or with the home-made programs. In all cases, the initial energy minimization is required to achieve the equilibrium state of the system.

For 2D materials, it is very important to take into account that they are one-atom-thick and the real thickness cannot be easily defined, which has been discussed previously [301, 307]. Thus, stress and elastic moduli of 2D structures are reported in force per unit length (N/m) rather than force per unit area (N/m^2 or Pa).

4.1.2 *Boundary Conditions*

Any system is composed of a finite number of atoms and limited by a surface; however, the surface atoms behave differently from the atoms in the bulk. What would happen when atoms from the surface move or detach from the structure?

Most of the simulations are conducted for parts of systems of much larger sizes, for example, to study defects in macroscopic systems. In this case, the modeled system is only a part of a large system containing the defect, and the atoms on the boundary of the modeled system belong to the bulk of the crystal in the physical system. If these atoms are considered free, then the influence of the surface on the state of the system will be significant. Therefore, to model the behavior of macroscopic systems, it is necessary to introduce special conditions on the atoms at the boundary of the simulation cell, called boundary conditions. There are several types of boundary conditions.

The most commonly used are periodic boundary conditions (PBCs). The atoms are contained in a simulation cell (shown in gray in Fig. 4.3). This cell is repeated an infinite number of times by translation in all three directions, filling the entire space (shown in white in Fig. 4.3). In other words, if one atom is located with coordinates $\vec{r}$ in the simulation cell, this atom is considered to actually represent an infinite set of atoms located at the points $\vec{r} = l\vec{a} + m\vec{b} + n\vec{c}$, $(l, m, n = -\infty, \infty)$, where l, m, n are integers; $\vec{a}, \vec{b}, \vec{c}$ correspond to the sides of the simulation cell. The key is the assumption that each i-th atom in the simulation cell interacts not only with the other atoms in the same cell, but also with their images in neighboring cells. That is, interactions can occur across the boundaries of a simulation cell. This leads to the fact that (a) we actually exclude the influence of the surface on the system, and (b) the position of the cell boundaries relative to the atoms does not matter. During the simulation, any atom of the system can leave the simulation cell. In this case, a periodic image of this atom enters the cell through the opposite boundary, so that the number of atoms in the cell does not change.

The other type of boundary conditions is the fixed boundary, which can be considered as a rigid wall. Atoms interacting with this wall will collide with the boundary and come back to the simulation cell. Such boundary

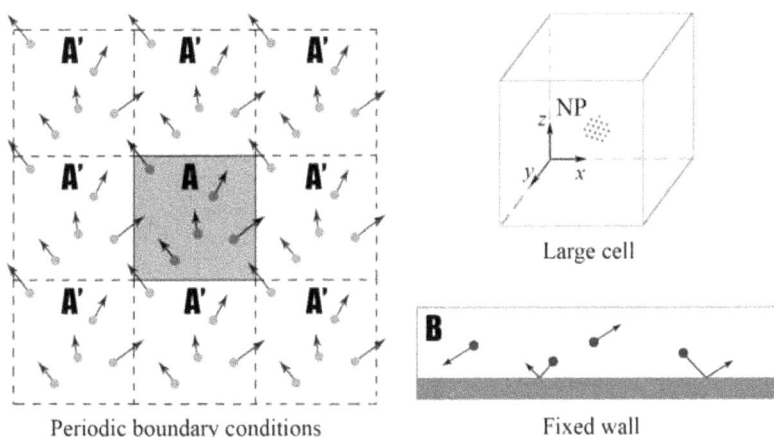

Periodic boundary conditions   Fixed wall

Fig. 4.3. Boundary conditions.

conditions can be applied, for example, when nanoindentation of the system is considered.

If we need to simulate a cluster of atoms, then there is no need to do anything special with atoms on the surface, and can be considered as shown in Fig. 4.3. For example, if we need to analyze the melting of the nanocluster, then we should give atoms an empty space for movement. We can choose a large simulation cell and allow the nanocluster to freely move inside.

Figure 4.4 presents the example of right and wrong boundary conditions for graphene. As can be seen, size of the simulation cell (L_x, L_y, L_z) together with the applied boundary conditions are very important for final simulation results. For example, if the size of the simulation cell along z-axis is too small, graphite-like structure will be simulated (imaginary layers will interact with the graphene in the simulation cell) and results will be very different from a one-layer structure. If one needs to simulate a single graphene layer, then PBC should be applied along x- and y-axes, while along z-axis L_z should be much larger.

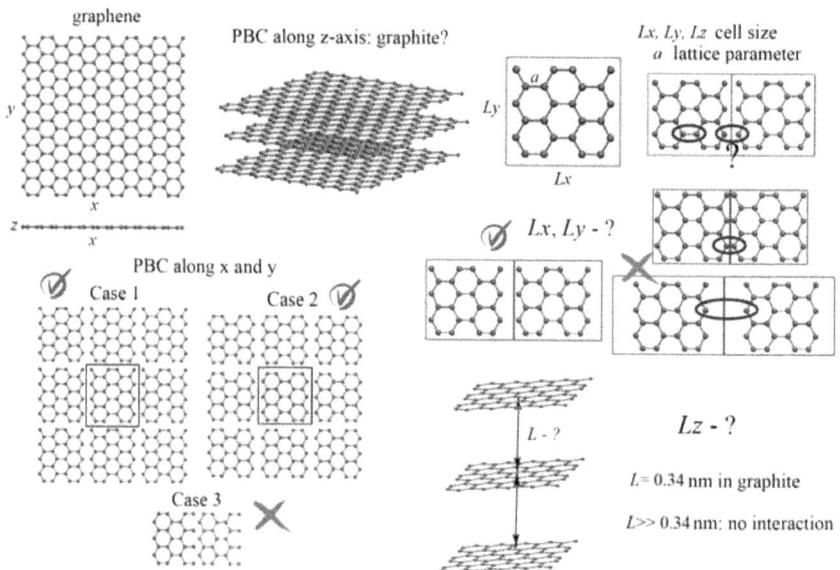

Fig. 4.4.   Example of right and wrong boundary conditions for graphene. PBC - periodic boundary conditions.

Size of the simulation cell along x- and y-axes is even more important. In Fig. 4.4, three cases on how to choose the simulation cell are presented. PBC means that we will have eight imaginary simulation cells and graphene can be considered as infinite. Simulation cells can be chosen in very different ways, but what is more important is that the lattice should be preserved when imaginary cells are under consideration. Case 3 is wrong, because on the borders of the simulation cell atoms cannot properly bond to each other.

Size of the simulation cell is very important when taking into account that the imaginary cell should be placed at the right position in accordance with the lattice order. Lattice parameter and atomic positions should be the same to the structure on the borders as well. Two wrong examples are presented in Fig. 4.4: simulation cell is too small and atoms on the borders are placed too close; and simulation cell is too big and atoms in the imaginary simulation cell will not interact with the main cell so it will be an empty space between cells.

Choice of the boundary conditions and proper size of the simulation cell will considerably affect the obtained results and their physical meaning.

4.1.3 *Interatomic Potentials*

When conducting MD studies, the interatomic interaction potential plays a critical role, since it determines how the interaction of atoms in the system will be described. Despite the fact that the interatomic interaction potentials described above are well tested and actively used in calculations, there are certain areas of applicability for certain potentials.

Bond order potentials (BOP) are a class of empirical (analytical) interatomic potentials used in MD. Other potentials that can be included in this class are the Tersoff potential [264], the Brenner potential [32], Adaptive Intermolecular Reactive Empirical Bond Order (AIREBO) [262], the Finnis-Sinclair potentials [73], or the ReaxFF (reactive force field potential) [273]. The main advantage is that they can describe several different bond states of an atom with the same parameters, and can therefore describe chemical reactions relatively reliably. The potentials were developed largely independently of each other, but share the common idea that the strength of a chemical bond depends on the environment, including the number of bonds and possibly the magnitude of bond angles and

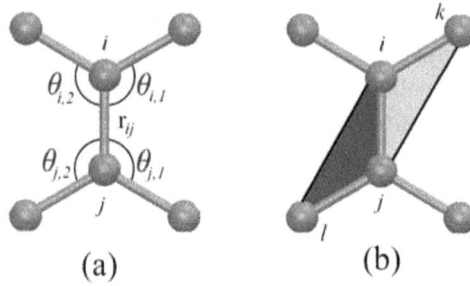

Fig. 4.5.   Part of the hexagonal lattice.

bond lengths. These potentials, in common, can be written as:

$$V_{ij}(r_{ij}) = V_{rep}(r_{ij}) + b_{ijk}V_{att}(r_{ij}),$$

where V_{rep} and V_{att} respectively refer to the repulsive and attractive parts.

Figure 4.5a presents a part of the typical hexagonal lattice (graphene, for example). Here, bond length r_{ij} is the bond between two carbon atoms, with indices i and j, and $\theta_{i,1}$, $\theta_{i,2}$ and $\theta_{j,1}$, $\theta_{j,2}$ the angles between the valence bonds.

4.1.3.1   *Brenner Potential*

The bond stretching is often represented by a simple harmonic function, which indicates that the bond cannot be broken and is only suitable for systems with bond stretching around its equilibrium states [273]. Then the interaction energy is

$$U_{ij} = V_R(r_{ij}) - \frac{1}{2}(B_{ij} + B_{ji})V_A(r_{ij}),$$

where the repulsive part is

$$V_R(r_{ij}) = \frac{D}{S-1}exp[-\sqrt{2S}\beta(r-r_0)],$$

and attractive part is

$$V_A(r_{ij}) = \frac{DS}{S-1}exp[-\sqrt{2/S}\beta(r-r_0)],$$

where the parameters are $r_0 = 1.39$ Å, $D = 6.0$ eV, $\beta = 2.1$ Å^{-1}, and $S = 1.22$ [245].

The angle potential is a three-body interaction, describing the angular vibrational motion occurring between three adjacent covalently bonded atoms. The term B_{ij} is the function of θ angles and can be defined as

$$B_{ij} = [1 + G(\theta_{i,1}) + G(\theta_{i,2})]^{-\delta},$$
$$B_{ji} = [1 + G(\theta_{j,1}) + G(\theta_{j,2})]^{-\delta},$$

and can be defined through

$$G(\theta) = a_0 \left[1 + \frac{c_0^2}{d_0^2} - \frac{c_0^2}{d_0^2 + (1 + \cos\theta)^2} \right],$$

where $a_0 = 0.00020812$, $c_0 = 330$, $d_0 = 3.5$, and $\delta = 0.5$.

The potential is written as a simple pair potential depending on the distance between the two atoms r_{ij}, but the bond strength depends on the environment of the i-th atom through the term b_{ijk}. Accordingly, the energy can be written as:

$$V_{ij}(r_{ij}) = V_{pair}(r_{ij}) - D\sqrt{\rho_i},$$

where ρ_i is the electron density near the i-th atom.

Tersoff and Brenner potentials are typical BOP potentials, developed for the simulation of covalent crystals like silicone or diamond. Potential energy of the i-th atom in the potential field from all the other particles in the system can be written as

$$V_i = \sum_{i \neq j} V_{ij}.$$

For Tersoff, interaction between i and j atoms can be considered as pair interatomic interaction and thus can be written as a modified isotropic Morse potential

$$V_{ij} = Ae^{-2\lambda r_{ij}} - Be^{-\lambda r_{ij}},$$

where $Ae^{-2\lambda r_{ij}}$ is a repulsive part and $Be^{-\lambda r_{ij}}$ is an attractive part. The transition from the isotropic Morse potential to the anisotropic Tersoff potential can be achieved by assuming that the coefficient B depends on the position of all neighboring particles:

$$B = B_{ij} = B_0 exp\left(\frac{-z_{ij}}{b}\right),$$

where

$$z_{ij} = \sum_{k \neq i,j} \left(\frac{e^{-r_{ij}}}{e^{-r_{ik}}} \right) \left(c + e^{-d \cos ijk} \right).$$

Thus, the weakening of the interatomic bond caused by the electron density pulling of neighboring particles is taken into account.

The interatomic potential is an exponentially decreasing function of the distance, which means that the potential is short-range. Taking into account the short-range nature, the summation over j and k can only be carried out over particles located at a distance less than a cut-off r_c, which significantly reduces the computational complexity of the algorithm. The parameters of the Tersoff potential are selected so as to satisfy a number of experimental facts.

4.1.3.2 *AIREBO Potential*

AIREBO potential is to describe the interatomic interactions between carbon atoms

$$U_{C-C} = \frac{1}{2} \sum_i \sum_{i \neq j} \left[U_{ij}^{REBO} + U_{ij}^{LJ} + \sum_{k \neq i,j} \sum_{l \neq i,j,k} U_{kijl}^{TORSION} \right], \quad (4.1)$$

where U_{ij}^{REBO} is the hydrocarbon REBO potential developed in [32], U_{ij}^{LJ} term adds longer-ranged interactions using a form similar to the standard Lennard-Jones (LJ) potential, and $U_{kijl}^{TORSION}$ describes various dihedral angle preferences in hydrocarbon configurations. This potential is very famous for the study of different carbon structures and their properties.

AIREBO potential is based on Brenner or REBO potential, but it was modified to represent more complex behavior of the covalent system. Two additional terms were added to the pair interaction: torsion and van der Waals interaction. The dihedral (torsion) potential is a four-body torsion angle, describing the angular spring between the planes formed by the first three and last three atoms of a consecutively bonded quadruple of atoms as shown in Fig. 4.5b, which constrains the rotation around a bond. Dihedral plays a crucial role in determining the structure and stability of a molecule. Energy of the torsion interaction for torsion angle between k, i, j, l atoms is written as

$$E_{kijl}^{tors} = \omega_{ki}(r_{ki})\omega_{ij}(r_{ij})\omega_{jl}(r_{jl})V^{tors}(\omega_{kijl}),$$

where torsion interaction is described by

$$V^{tors}(\omega_{kijl}) = \frac{256}{405}\varepsilon_{kijl}\cos^{10}(\omega_{kijl}/2) - \frac{1}{10}\varepsilon_{kijl}.$$

The van der Waals interaction exists between any two non-bonded atoms arising from the balance between attractive and repulsive potentials. To describe the van der Waals interactions, a simple LJ pair potential is used:

$$V_{ij}^{LJ} = 4\varepsilon_{ij}\left(\left(\frac{\sigma_{ij}}{r_{ij}}\right)^{12} - \left(\frac{\sigma_{ij}}{r_{ij}}\right)^{6}\right).$$

This part can be turned on/off depending on the goals of the simulation.

4.1.3.3 *Set of the Interatomic Potentials*

Potential energy

$$P = U_1 + U_2 + U_3 + U_4 + U_5,$$

defines the deformation of three covalent bonds, six valence angles, and 12 angles between four planes in a graphene lattice as presented in Fig. 4.6. The energy of valence angles can be found as

$$U_1(\vec{r}_1, \vec{r}_2) = \varepsilon_1\left[e^{-\alpha_0(\rho - \rho_0)} - 1\right]^2, \quad \rho = \vec{r}_2 - \vec{r}_1,$$

where parameter $\varepsilon_1 = 4.9632$.

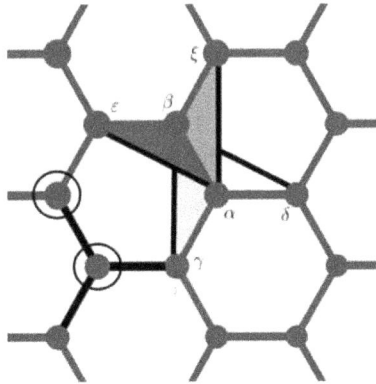

Fig. 4.6. Part of the hexagonal lattice with valence bonds and angles.

Changes of the valence angles can be defined as

$$U_2(\vec{r}_1, \vec{r}_2, \vec{r}_3) = \varepsilon_2(\cos\varphi - \cos\varphi_0)^2,$$

$$\cos\varphi = (\vec{r}_3 - \vec{r}_2, \vec{r}_1 - \vec{r}_2)(|\vec{r}_3 - \vec{r}_2||\vec{r}_2 - \vec{r}_1|),$$

where $\varphi = \pi/3$ – the equilibrium angle and parameter $\varepsilon = 1.3143$.

For each covalent bond, we consider four planes, labelled in Fig. 4.6 $\alpha\beta\gamma$, $\alpha\beta\delta$, $\alpha\beta\varepsilon$ and $\alpha\beta\xi$ and the energy associated with the six angles between these planes is calculated as

$$U_i(\vec{r}_1, \vec{r}_2, \vec{r}_3, \vec{r}_4) = \varepsilon_i(1 - \cos\phi), \quad i = 3, 4, 5, \quad \cos\phi = (\vec{v}_1, \vec{v}_2)/(|\vec{v}_1|, |\vec{v}_2|),$$

$$\vec{v}_1 = (\vec{u}_2 - \vec{u}_1) \times (\vec{u}_3 - \vec{u}_2), \quad \vec{v}_2 = (\vec{u}_3 - \vec{u}_2) \times (\vec{u}_3 - \vec{u}_4),$$

where $\vec{u}_i$ is the translation vector.

Energy U_3 describes the deformation of $\gamma\alpha\beta\delta$ and $\varepsilon\beta\alpha\xi$ angles, energy U_4 describes the deformation of $\gamma\alpha\beta\xi$ and $\varepsilon\beta\alpha\delta$ angles, and energy U_5 describes the deformation of $\gamma\alpha\beta\varepsilon$ and $\delta\alpha\beta\xi$ angles. Parameter ε_4 is close to ε_3, and $\varepsilon_5 \ll \varepsilon_4$. Therefore, we further take values of $\varepsilon_3 = \varepsilon_4 = 0.499$ eV and assume $\varepsilon_5 = 0$.

4.1.4   *Choice of the Interatomic Potential*

4.1.4.1   *Example for Graphene*

Let us consider the difference in the results obtained for sound velocities for graphene, simulated using two different potentials: Brenner potential and the set of interatomic potentials developed by A. V. Savin. Note that these potentials even give different equilibrium lattice constants: the Brenner potential gives $r_0 = 1.4505$ Å, while the standard set of interatomic potentials gives $r_0 = 1.418$ Å.

Figure 4.7 presents the sound velocities and dispersion curves of unstrained graphene, obtained with the Brenner potential (a) and set of potentials developed by A. V. Savin (b). A comparison of the dispersion curves of undeformed graphene (c) shows the difference in the results obtained by different potentials. The phonon spectrum of graphene includes three acoustic and three optical branches. The acoustic branches with the highest (LA) and average (TA) frequencies correspond to longitudinal and transverse waves in the graphene plane. The acoustic branch with the lowest frequency (ZA) corresponds to transverse waves outside the graphene plane. Brenner potential gives higher values than the set of the interatomic potentials, although no serious qualitative differences were observed. The

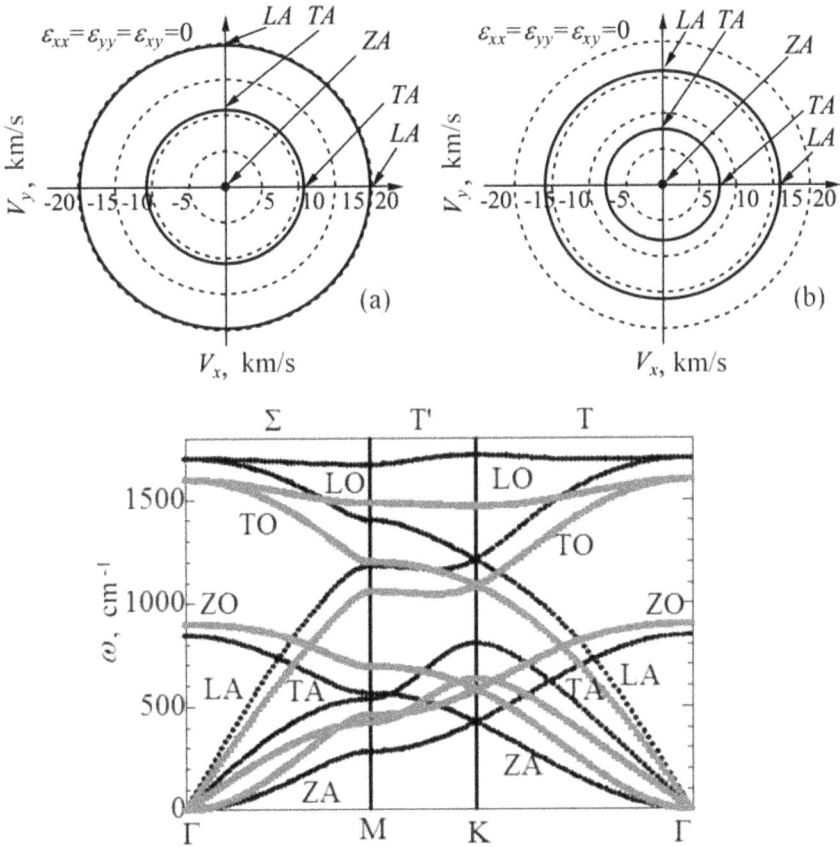

Fig. 4.7. Sound velocities obtained with the Brenner potential (a) and set of potentials developed by A. V. Savin (b).

comparison of the obtained values with experimental data, as well as with the results of other calculations, shows that the Brenner potential overestimates the frequency values. Therefore, when calculating dispersion curves and any other characteristics that can be calculated based on these data, the set of interatomic potentials developed by Savin should be used.

In the undeformed state, graphene is isotropic and, therefore, the sound velocity does not depend on the propagation direction and is equal to 19.7 (15.9) km/s for the phonon branch LA and 10.7 (7.76) km/s for the phonon branch TA, in the calculation with the Brenner potential (a set of interatomic potentials developed by A. V. Savin). In both calculations, the ZA wave has zero sound speed, since the bending rigidity of undeformed graphene is zero. Despite the difference in the values of sound velocities obtained with different potential functions, the qualitative agreement can be seen.

In experiments on inelastic X-ray scattering, it was found that the sound velocity in graphene is 22 km/s for longitudinal and 14 km/s for transverse waves in the plane, and the sound velocity for longitudinal waves in the plane, obtained in Raman scattering experiments, is 20 km/s. Thus, the Brenner potential with a modified set of parameters gives a better description of the sound velocities in undeformed graphene than the set of potentials of Savin. The deviation of the sound velocities obtained with the standard set of interatomic potentials can be explained by the difference in the parametric values.

Figure 4.8 presents the orientation dependence of the sound velocities for graphene under tension along the armchair direction. Again, there is a

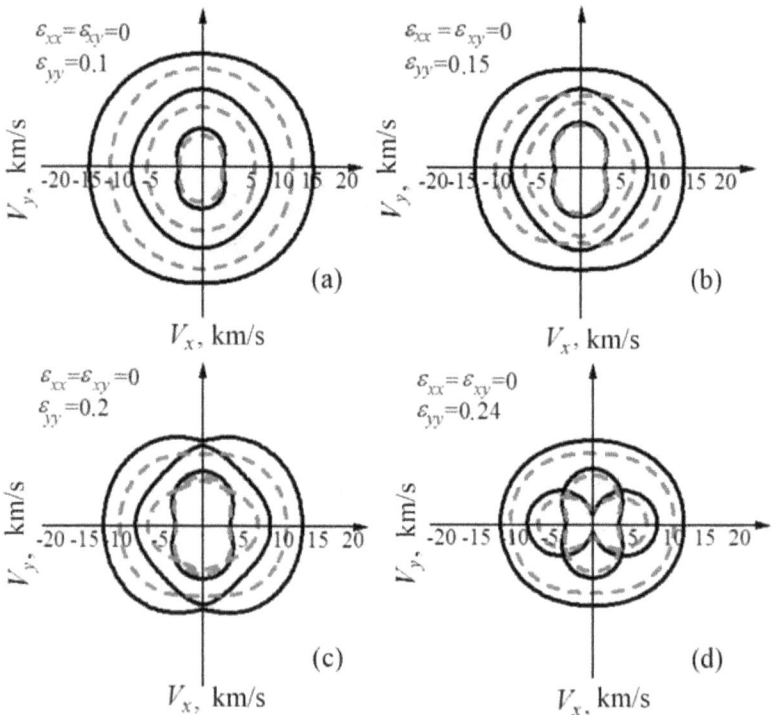

Fig. 4.8. Sound velocities at tension along the armchair direction, obtained with the Brenner potential (dashed grey lines) and set of interatomic potentials developed by Savin (solid black lines): (a) $\varepsilon_{yy} = 0.1$, (b) $\varepsilon_{yy} = 0.15$, (c) $\varepsilon_{yy} = 0.2$, (d) $\varepsilon_{yy} = 0.24$.

difference in the values obtained by different potentials, but qualitatively sound velocities change in similar ways.

4.1.4.2 *Example for Graphene/Metal System*

For the simulation of more complex systems composed of different atoms, hybrid interatomic potentials can be used. For example, to simulate interactions in the graphene/metal system, carbon-carbon (C-C), metal-carbon (Me-C), and metal-metal (Me-Me) interactions should be taken into account. It is possible to describe all interactions (C-C, C-Me, and Me-Me) using just one complex potential, for example, with the modified BOP developed to describe the interaction in the Cu-C system [308], BOP potential for Ti_3AlC_2 [76, 210], ReaxFF for Ni-C system [21] or third-generation charge-optimized many-body (COMB3) interatomic potential for Al-C system [300]. The other way is to reproduce different interactions separately: BOP for C-C, embedded atom method (EAM) for Me-Me, and pair Morse or LJ interatomic potential for Me-C.

The Tersoff [264], Brenner [32] or AIREBO [262] potential can be effectively used to study amorphous carbon, graphite, diamond or carbon polymorphs with high accuracy. To describe the interaction between metal atoms, EAM potential is most widely used. To simulate carbon/metal van der Waals interaction, the simple pair LJ and Morse potentials can be used. In Table 4.1, the LJ potential parameters, as well as the cutoff radius (r_{cut}) for Al-C, Ni-C, Ti-C, and Cu-C are presented.

The Morse potential is written in the following form:

$$E_{Me-C}(r) = D_e[(1 - e^{-\beta(r-R_e)})^2 - 1], \qquad (4.2)$$

where D_e is the binding energy, R_e equilibrium distance and β – potential parameter characterizing the bond strength.

Table 4.2 presents the Morse potential parameters for Al-C, Ni-C, Ti-C and Cu-C.

Table 4.1. Parameters of LJ potential for graphene/Me interactions.

System	ε, eV	σ, Å	r_{cut}, Å	Ref.
Al-C	0.03438	3.01	10.2	[49]
Ni-C	0.023049	2.852	10.0	[294]
Ti-C	0.006535	3.6	10.8	[195]
Cu-C	0.01996	3.225	8.0625	[108]

Table 4.2. Parameters of Morse potential for graphene/Me interactions.

System	D_e, eV	α, 1/Å	R_e, Å	r_{cut}, Å	Ref.
Al-C	0.196	4.0170	3.450	4.0	[119]
Ni-C	0.433	3.2441	2.316	4.5	[77]
Ti-C	0.982	2.283	1.892	7.4	[312]
Cu-C	−0.10	1.70	2.22	6.5	[278]

The EAM potential is a well-known method to describe Me-Me interactions. EAM potential parameterized by Mendelev will be used to describe the Al-Al [166], Cu-Cu [165], Ti-Ti [168] and Ni-Ni [167] interactions.

Four interatomic potentials AIREBO-LJ-EAM, AIREBO-Morse-EAM, Tersoff and BOP are examined for the graphene/Me system. Further, special notations will be introduced: results obtained with the hybrid AIREBO-LJ-EAM potential will be called *LJ*, obtained with the hybrid AIREBO-Morse-EAM potential will be called *Morse* and obtained with complex potentials will be called *Tersoff* or *BOP*.

Let us consider the interaction of graphene with one Me (Ni, Cu, Al, Ti) atom. It is found that the total potential energy per atom for all systems is a constant value, indicating that the system reached an equilibrium state.

For graphene interacting with Al atom at room temperature, the equilibrium state is reached at an energy of about −7.110 eV/atom for any considered potential. For the structure simulated with LJ potential, the Al atom moves over the graphene surface, while for the structure simulated with Tersoff potential, the Al atom remains in the same position during all exposure time. For both structures, the equilibrium distance between the Al atom and graphene remains about 3.2 Å (the equilibrium distance is 3.36–3.41 Å). When using the Morse potential, the Al atom detaches from graphene and freely moves in the computational cell.

For exposure of Cu atom and graphene at 300 K, the LJ and Morse potentials show almost the same result, with equilibrium potential energy of about −7.110 eV/atom, in contrast to the BOP potential, where equilibrium state is reached at a lower potential energy of −7.156 eV/atom. Visual analysis of the structures confirms that for LJ or Morse, Cu atom detaches from the graphene surface and freely moves in the simulation cell. The BOP potential shows the opposite result: the Cu atom is attracted by graphene and does not change its position during exposure. Moreover, graphene simulated with BOP potential remains flat, which contradicts other results.

For Ni atom interacting with graphene, equilibrium potential energy is about -7.113 eV/atom and in all cases (LJ, Morse and ReaxFF potentials) is attached to the graphene surface. The equilibrium distance between graphene and Ni with the LJ potential is 2.8 Å, and with the Morse potential is 2.4 Å, while the optimal distance between graphene and Ni is 2.018–2.1 Å.

For the graphene/Ti system, modeling using LJ, Morse, and Tersoff potentials results in the same potential energy value of -7.115 eV/atom. However, during exposure at 300 K, different structural states are observed. For the graphene/Ti system simulated using LJ potential, the Ti atom moves away from graphene. For the system simulated using Morse and Tersoff potentials, it remains on the graphene surface. The equilibrium distance between Ti atom and graphene surface is 1.8 and 2.5 Å for the Morse and Tersoff potentials, respectively (the equilibrium distance between an adsorbed Ti atom and the graphene surface is 2.34 Å).

The analysis of the interaction of graphene with spherical metal nanoparticles at the same exposure temperature (300 K) can give qualitatively different results, because the shape of graphene and the metal is of high importnace. Figure 4.9 shows the potential energy per atom versus the exposure time at room temperature for different graphene/Me systems, obtained using different interatomic potentials.

It can be seen from Fig. 4.9 that for GR/Al with LJ, Morse or Tersoff potential, the equilibrium potential energy is the same and equal to -5.70 eV/atom. LJ potential gives strong interaction between graphene and Al nanoparticle – graphene covers Al, while Morse and Tersoff potentials show similar results – Al is attracted to graphene, but not covered. Since the interaction energy between Al and graphene is rather weak [87], the Morse and Tersoff potentials demonstrate better physical representation.

For the graphene/Cu system, the equilibrium potential energy is -4.41 eV/atom (for LJ and Morse), and -7.37 eV/atom (for BOP). For Morse and LJ potentials, the Cu nanoparticle moves over the graphene surface during exposure, while for BOP potential, a fairly strong interaction with graphene is observed. However, there should be a rather weak bonding between Cu and graphene [87, 97, 291]. Hence, we can conclude that the LJ and Morse potentials are more suitable for the simulation of the interaction in the graphene/Cu system.

For the graphene/Ni system, the potential energy for two different potentials (LJ and Morse) are approximately in the same range of values, and almost the same behavior of structures is observed during exposure at 300 K: the Ni nanoparticle is attached to graphene and covered by it. In this

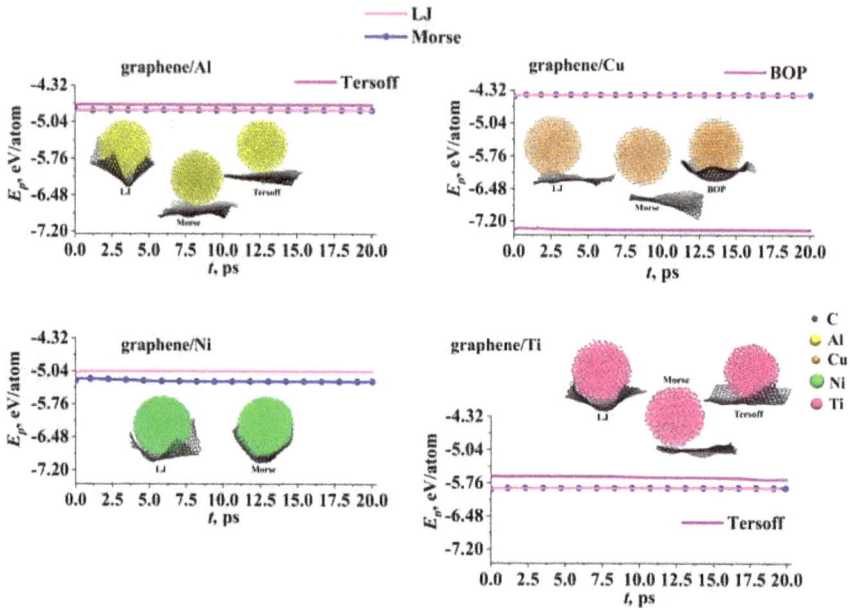

Fig. 4.9. Potential energy as a function of exposure time obtained using different potentials for a structure consisting of graphene and a metal nanoparticle. Snapshots of the structure after exposure for 20 ps.

case, both potentials are suitable for the simulation of the graphene/Ni system.

The equilibrium state for the graphene/Ti system is reached at -5.94 eV/atom and -5.88 eV/atom for Tersoff (Morse) and LJ potentials, respectively. For the LJ and Tersoff potential, the graphene edges bend and stretch towards the Ti nanoparticle. After exposure at 300 K for 20 ps, graphene completely covers Ti. The Morse potential shows a weaker interaction between graphene and Ti and graphene remains almost flat. The LJ potential shows the strongest interaction between graphene and Ti.

The interaction in a more complex graphene/Me composite precursor using different interatomic potentials again will differ from the interaction of Me atom or Me nanoparticles with graphene. Let us considre the system of graphene flakes filled with Me nanoparticles.

For the graphene/Al system (see Fig. 4.10), LJ and Morse potentials give the same values of the potential energy. For the Tersoff potential, a sharp decrease of the potential energy up to -6.87 eV/atom is observed.

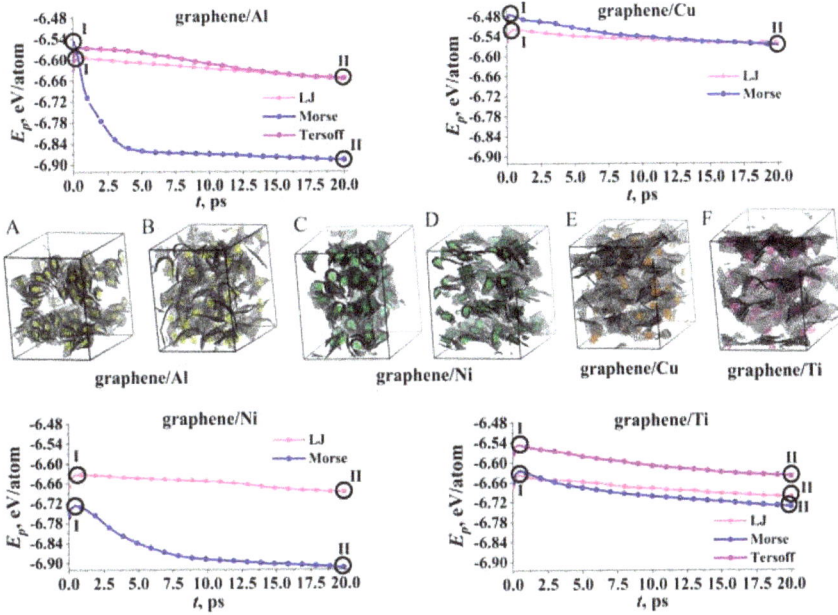

Fig. 4.10. Potential energy as a function of exposure time obtained using different potentials for the structure of crumpled graphene filled with Al/Ni/Cu/Ti nanoparticles. Snapshots after exposure (II).

The potential energy at point II in Fig. 4.10 is -6.89 eV/atom for the Tersoff potential, which is 0.25 eV/atom lower than for the LJ and Morse potentials. Such a difference in potential energy during exposure is associated with different transformations of the structure.

Snapshot B corresponds to II for Tersoff or Morse, while snapshot A is for LJ. The LJ potential shows a strong interaction between graphene flake and Al nanoparticles. With exposure individual elements of the composite become more flattened due to the uniform distribution of Al nanoparticles over the inner surface of graphene. For Tersoff or Morse during exposure, graphene flakes open and Al nanoparticles leave the internal cavity of graphene (snapshot B). The Morse and Tersoff potentials show a weak interaction between graphene and Al nanoparticles. Instead, they interact more easily with each other than with graphene, which leads to their coagulation into large conglomerates. Since the bond between Al and graphene is indeed quite weak [87], then the nature of the interaction between Al nanoparticles and graphene, described by the Morse and Tersoff potentials, best demonstrates the physical behavior in this system.

For the graphene/Cu system, the potential energy obtained using the LJ and Morse potentials almost coincides. BOP potential is not analyzed because it demonstrates wrong behavior for graphene and one metal atom. The behavior of the structures during exposure at 300 K is the same (snapshot E) for both potentials. Graphene flakes filled with Cu nanoparticles almost immediately open and become flat. In this case, Cu nanoparticles move from the graphene surface but do not leave the structure and do not coagulate with each other, as in the case of Al nanoparticles.

The potential energy curves for the graphene/Ni system are different for LJ and Morse potentials (see Fig. 4.10). The Morse potential shows a sharp decrease in the potential energy of the system at 10 ps, and a further increase in the exposure time leads to a slight decrease in the value of E_p. An analysis of the structures during exposure shows that when the LJ potential is used, Ni nanoparticles retain the crystal structure of fcc Ni. When the Morse potential is used, the crystalline order is destroyed due to the strong interaction between Ni and graphene. This leads to the splitting of nickel nanoparticles into smaller particles deposited on the graphene surface. When using the LJ potential, the strong interaction of individual elements is observed (snapshot C).

For the graphene/Ti system, the potential energy curves are very similar as well as the changes in the structure during the simulation. At the beginning of exposure, the graphene flakes tend to open up, and Ti nanoparticles are attracted to the graphene surface. An analysis of the obtained simulation data showed that all three potentials used to describe the graphene/Ti system led to qualitatively the same result.

The analysis of the exposure of graphene/metal systems at a temperature of 300 K using different simulation potentials showed that the same interaction potential for structures with different configurations (graphene with one metal atom, graphene with metal nanoparticles, and 3D graphene/Me composite precursor) can demonstrate different changes in the structure and interaction of elements in the system. For convenience, Table 4.3 shows all the investigated interaction potentials for the graphene/Me systems.

For example, for graphene with one Al atom simulated with Morse, the Al atom moves away from the graphene surface, although the equilibrium distance between the Al atom and graphene is 3.36–3.41 Å [67, 87, 300]. Two other potentials (LJ and Tersoff) better describe such a system since the equilibrium distance in the interaction of an Al atom and graphene is about 3.2 Å in both cases. However, when considering Al nanoparticles

Table 4.3. Summary table on the possible application
of interaction potentials for graphene/Me system.

Potential	Atom	Nanoparticle	Composite
graphene/Al			
LJ	+	−	−
Morse	−	+	+
Tersoff	+	+	+
graphene/Cu			
LJ	−	+	+
Morse	−	+	+
BOP	+	−	−
graphene/Ni			
LJ	−	+	+
Morse	+	+	+
graphene/Ti			
LJ	−	+	+
Morse	+	−	−
Tersoff	+	+	+

on the surface of graphene or 3D graphene/Al composite precursor, better physical results are obtained using the Morse and Tersoff potentials. These potentials show weak bonding between Al and graphene, which does not lead to adhesion of graphene to Al, as when using the LJ potential. This contradicts the literature [87, 97, 291], therefore the LJ potential cannot be used to study graphene/Al composites.

For the graphene/Cu system, when using the LJ and Morse potentials for a graphene system with one Cu atom, it moves from the graphene surface at a distance significantly larger than the equilibrium distance (3.26 Å [87, 291, 308]). This is not observed for the BOP potential, and the equilibrium distance between Cu atom and graphene is 3.2 Å. Therefore, only this potential can be used to study such systems. However, when moving to more complex configurations, such as graphene with Cu nanoparticles and graphene/Cu composite precursor, the BOP complex potential demonstrates a fairly strong interaction between graphene and copper, which is not typical for this system [87, 97, 291]. When using the LJ and Morse potentials, a weaker interaction between graphene and Cu nanoparticles is observed.

For graphene/Ni, the results obtained by simulation using the LJ and Morse potentials are approximately the same for all configurations. However, when using the LJ potential, graphene flakes with Ni nanoparticles inside begin to combine in the exposure process. This is not observed when modeling with the Morse potential. In addition, during the exposure of graphene with one Ni atom, the equilibrium distance between Ni and the graphene surface is close to the literature values (2.018–2.1 Å [87, 95, 291]) when using the Morse potential (2.4 Å). However, both potentials describe well the strong interaction between Ni and graphene, which is typical for this system [87, 97, 291]. Therefore, both potentials can be used to simulate graphene/Ni composite.

For the graphene/Ti system, for graphene with one Ti atom simulated using the Tersoff potential, the equilibrium distance between Ti and the graphene surface (2.5 Å) is close to that from literature (2.34 and 2.17 Å, respectively [229, 298]). When using the LJ potential during exposure, the Ti atom leaves the graphene surface, which is not physically right, since a fairly strong bond is observed for graphene and Ti [229, 298]. Therefore, the LJ potential is not suitable for the simulation of a graphene system with one Ti atom. However, when a more complex system is simulated, a stronger bond between graphene and Ti nanoparticles is observed in the structure obtained using the LJ potential. The weakest interaction is observed when using the Morse potential. For the graphene/Ti composite precursor, no significant differences in the structure are observed for all potentials used. However, when using the Morse potential, many Ti nanoparticles are detached from the graphene surface, which leads to their coagulation with each other or to interaction with other graphene flakes. This indicates a weaker bond between Ti nanoparticles and graphene.

4.2   Elastic Constants

Figure 4.11 presents the simulation cells for diamane with thickness h. PBCs are applied along x-, y- and z-axes; however, the size of the simulation cell normal to the plane of the 2D layer is much larger than the thickness of the 2D layer to avoid the van der Waals interaction between imaginary layers of the periodic cell. This methodology is commonly used for the simulation of similar structures [235, 269].

In Fig. 4.11, the stress-strain curve (blue) during uniaxial tension for D-AA-H is presented. For comparison, linear representation is shown by a red dashed line. It can be seen that linear elastic deformation is observed up to $\varepsilon = 0.015$. Thus, beyond that only strain lower than 1% is applied, which is also in agreement with previous results [197].

For the case of diamane, since its dimension in the thickness direction h is much lower than the size of the simulation box H, the stress in diamane

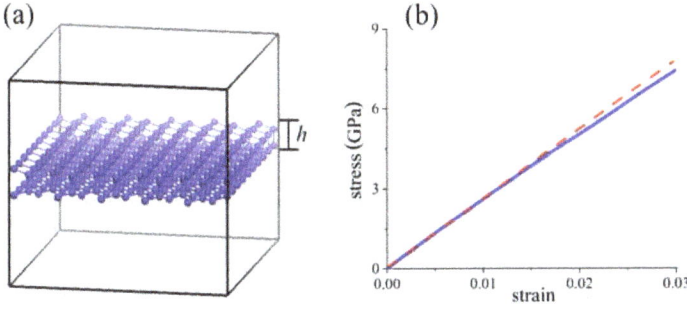

Fig. 4.11. (a) Schematic of the simulation box for AA-stacked diamane (D-AA). (b) Stress-strain curve during uniaxial tension of D-AA (blue color). The red dashed line shows the linear representation.

can be obtained from the stress applied to the simulation box σ_{box} [269], simply from the geometric parameters as:

$$\sigma = \frac{\sigma_{box} H}{h}. \tag{4.3}$$

According to Hooke's law, at small strains, the stress components σ_{ij} are directly proportional to the strain components ε_{kl} and have the form:

$$\sigma_{ij} = c_{ijkl} \varepsilon_{kl} \tag{4.4}$$

where c_{ijkl} is the fourth-order tensor of stiffness constant. In this case, Hooke's law for materials is written as a matrix:

$$
\begin{pmatrix}
\sigma_{xx} \\
\sigma_{yy} \\
\sigma_{zz} \\
\sigma_{yz} \\
\sigma_{xz} \\
\sigma_{xy}
\end{pmatrix}
=
\begin{pmatrix}
c_{11} & c_{12} & c_{13} & c_{14} & c_{15} & c_{16} \\
c_{21} & c_{22} & c_{23} & c_{24} & c_{25} & c_{26} \\
c_{31} & c_{32} & c_{33} & c_{34} & c_{35} & c_{36} \\
c_{41} & c_{42} & c_{43} & c_{44} & c_{45} & c_{46} \\
c_{51} & c_{52} & c_{53} & c_{54} & c_{55} & c_{56} \\
c_{61} & c_{62} & c_{63} & c_{64} & c_{65} & c_{66}
\end{pmatrix}
\begin{pmatrix}
\varepsilon_{xx} \\
\varepsilon_{yy} \\
\varepsilon_{zz} \\
\varepsilon_{yz} \\
\varepsilon_{xz} \\
\varepsilon_{xy}
\end{pmatrix}
\tag{4.5}
$$

where c_{ij} is a matrix of stiffness constants.

Note that after the relaxation of the system, it is still not at the global minimum of potential energy, which is the limitation of the MD simulation. A small applied strain can significantly affect the elastic constants obtained, especially for a 2D nanostructure. Therefore, to exclude computational errors, several numerical experiments should be performed for the preliminary uniaxial tension/compression. A uniaxial tensile strain of up to 0.1% is then applied to each of these structures and the stiffness constants are calculated.

To calculate the stiffness constants of all structures, a strain is applied to the simulation cell, and the corresponding stresses are calculated. The stiffness constants are calculated using the above equations. For 2D structures with atomic thickness, such as graphene (graphane, etc.), the exact value of the thickness cannot be readily determined, although it is necessary for the calculation of the elastic constants. Therefore, the elastic constants are measured in N/m rather than in GPa, as is common for 3D structures.

The equations presented below for calculating the coefficients of compliance and rigidity can be considered sufficient. On their basis, the thermodynamic stability of crystals can be analyzed and engineering elastic constants can be calculated. There is a slight difference for the calculation of elastic constants for 3D and 2D crystals, which also will be described further.

The number of stiffness and compliance coefficients for crystals of different anisptropy differs. For cubic anisptropy, three coefficients of elasticity are calculated, for hexagonal and trigonal crystals five coefficients of elasticity, for tetragonal crystals six coefficients of elasticity and for rhombic anisotropy nine coefficients of elasticity [178]. This number will be decreased for 2D structures [213, 214].

4.2.1 Cubic Crystals

For a cubic crystal it is enough to calculate only three compliance coefficients s_{11}, s_{12} and s_{44}; the expression takes the form of a matrix:

$$
\begin{pmatrix} \varepsilon_{xx} \\ \varepsilon_{yy} \\ \varepsilon_{zz} \\ \varepsilon_{yz} \\ \varepsilon_{xz} \\ \varepsilon_{xy} \end{pmatrix} = \begin{pmatrix} s_{11} & s_{12} & s_{12} & \cdot & \cdot & \cdot \\ s_{12} & s_{11} & s_{12} & \cdot & \cdot & \cdot \\ s_{12} & s_{12} & s_{11} & \cdot & \cdot & \cdot \\ \cdot & \cdot & \cdot & s_{44} & \cdot & \cdot \\ \cdot & \cdot & \cdot & \cdot & s_{44} & \cdot \\ \cdot & \cdot & \cdot & \cdot & \cdot & s_{44} \end{pmatrix} \begin{pmatrix} \sigma_{xx} \\ \sigma_{yy} \\ \sigma_{zz} \\ \sigma_{yz} \\ \sigma_{xz} \\ \sigma_{xy} \end{pmatrix}. \quad (4.6)
$$

Compliance constants can be calculated as:

$$s_{11} = \frac{\varepsilon_{xx}}{\sigma_{xx}}, \quad s_{12} = \frac{\varepsilon_{yy}}{\sigma_{xx}}, \quad s_{44} = \frac{\varepsilon_{xy}}{\sigma_{xy}}.$$

Stiffness constants are connected with the compliance coefficients as

$$c_{11} + c_{12} = \frac{1}{s_{11} - s_{12}}, \quad c_{11} - c_{12} = -\frac{s_{11}}{(s_{11} - s_{12})(s_{11} + 2s_{12})},$$

$$c_{44} = \frac{1}{s_{44}}.$$

4.2.2 Tetragonal, Hexagonal and Trigonal Crystals

Hook's law for tetragonal crystal:

$$
\begin{pmatrix}
\varepsilon_{xx} \\
\varepsilon_{yy} \\
\varepsilon_{zz} \\
\varepsilon_{yz} \\
\varepsilon_{xz} \\
\varepsilon_{xy}
\end{pmatrix}
=
\begin{pmatrix}
s_{11} & s_{12} & s_{13} & \cdot & \cdot & \cdot \\
s_{12} & s_{11} & s_{13} & \cdot & \cdot & \cdot \\
s_{12} & s_{13} & s_{33} & \cdot & \cdot & \cdot \\
\cdot & \cdot & \cdot & s_{44} & \cdot & \cdot \\
\cdot & \cdot & \cdot & \cdot & s_{44} & \cdot \\
\cdot & \cdot & \cdot & \cdot & \cdot & s_{66}
\end{pmatrix}
\begin{pmatrix}
\sigma_{xx} \\
\sigma_{yy} \\
\sigma_{zz} \\
\sigma_{yz} \\
\sigma_{xz} \\
\sigma_{xy}
\end{pmatrix}
\tag{4.7}
$$

where s_{11}, s_{12}, s_{13}, s_{33}, s_{44}, and s_{66} are compliance constants.

Tetragonal crystals have six independent compliance coefficients. For hexagonal crystals the number of independent compliance coefficients decreases to five: s_{11}, s_{12}, s_{13}, s_{33}, and s_{44} with additional relations $s_{66} = 2(s_{11} - s_{12})$.

In the case of trigonal symmetry, the compliance coefficients are calculated as for a hexagonal crystal. Here, the coefficient $s_{14} = \varepsilon_{xz}/\sigma_{xx}$ is added and s_{66} is not taken into account.

Based on Hooke's law, the compliance coefficients are calculated as:

$$s_{11} = \frac{\varepsilon_{xx}}{\sigma_{xx}}, \quad s_{12} = \frac{\varepsilon_{yy}}{\sigma_{xx}}, \quad s_{13} = \frac{\varepsilon_{zz}}{\sigma_{xx}}, \quad s_{33} = \frac{\varepsilon_{zz}}{\sigma_{zz}},$$

$$s_{44} = \frac{\varepsilon_{xy}}{\sigma_{xy}}, \quad s_{66} = \frac{\varepsilon_{xz}}{\sigma_{xz}}.$$

Stiffness constants can be calculated as:

$$c_{11} + c_{12} = \frac{s_{33}}{s}, \quad c_{11} - c_{12} = \frac{1}{(s_{11} - s_{12})}, \quad c_{13} = -\frac{s_{13}}{s},$$

$$c_{33} = \frac{s_{11} + s_{12}}{s}, \quad c_{44} = \frac{1}{s_{44}}, \quad c_{66} = \frac{1}{s_{66}},$$

where $s = s_{33}(s_{11} + s_{12}) - 2s_{13}^2 > 0$.

4.2.3   Rhombic Crystals

Hooke's law for rhombic crystal:

$$
\begin{pmatrix} \varepsilon_{xx} \\ \varepsilon_{yy} \\ \varepsilon_{zz} \\ \varepsilon_{yz} \\ \varepsilon_{xz} \\ \varepsilon_{xy} \end{pmatrix}
=
\begin{pmatrix}
s_{11} & s_{12} & s_{13} & \cdot & \cdot & \cdot \\
s_{12} & s_{22} & s_{23} & \cdot & \cdot & \cdot \\
s_{13} & s_{23} & s_{33} & \cdot & \cdot & \cdot \\
\cdot & \cdot & \cdot & s_{44} & \cdot & \cdot \\
\cdot & \cdot & \cdot & \cdot & s_{55} & \cdot \\
\cdot & \cdot & \cdot & \cdot & \cdot & s_{66}
\end{pmatrix}
\begin{pmatrix} \sigma_{xx} \\ \sigma_{yy} \\ \sigma_{zz} \\ \sigma_{yz} \\ \sigma_{xz} \\ \sigma_{xy} \end{pmatrix}
\tag{4.8}
$$

where s_{11}, s_{12}, s_{13}, s_{23}, s_{33}, s_{44}, s_{55}, and s_{66} are compliance constants.

For rhombic crystal, nine compliance and stiffness constants can be calculated. Based on Hooke's law for anisotropic bodies, the compliance coefficients are as follows:

$$
s_{11} = \frac{\varepsilon_{xx}}{\sigma_{xx}}, \quad s_{12} = \frac{\varepsilon_{yy}}{\sigma_{xx}}, \quad s_{13} = \frac{\varepsilon_{zz}}{\sigma_{xx}}, \quad s_{22} = \frac{\varepsilon_{yy}}{\sigma_{yy}},
$$

$$
s_{33} = \frac{\varepsilon_{zz}}{\sigma_{zz}}, \quad s_{23} = \frac{\varepsilon_{zz}}{\sigma_{yy}}, \quad s_{44} = \frac{\varepsilon_{xy}}{\sigma_{xy}}, \quad s_{55} = \frac{\varepsilon_{yz}}{\sigma_{yz}}, \quad s_{66} = \frac{\varepsilon_{xz}}{\sigma_{xz}}.
$$

Stiffness constants can be calculated as:

$$
c_{11} = \frac{s_{22}s_{33} - s_{23}^2}{s}, \quad c_{12} = \frac{s_{13}s_{23} - s_{12}s_{33}}{s}, \quad c_{13} = \frac{s_{12}s_{23} - s_{13}s_{22}}{s},
$$

$$
c_{22} = \frac{s_{11}s_{33} - s_{13}^2}{s}, \quad c_{23} = \frac{s_{12}s_{13} - s_{23}s_{11}}{s}, \quad c_{33} = \frac{s_{11}s_{22} - s_{12}^2}{s},
$$

$$
c_{44} = \frac{1}{s_{44}}, \quad c_{55} = \frac{1}{s_{55}}, \quad c_{66} = \frac{1}{s_{66}},
$$

where $s = (s_{11}s_{22} - s_{12}^2)s_{33} + 2s_{12}s_{13}s_{23} - s_{11}s_{23}^2 - s_{22}s_{13}^2$.

4.2.4   Hexagonal, Orthorhombic 2D Crystals

The above equations are usually applied to a 3D crystal, but structures under consideration are 2D and the thickness along z-axis is very low. Thus, the number of stiffness coefficients c_{ij} is reduced to three – c_{11}, c_{12}, c_{66}, and the stiffness matrix has the form

$$
||c_{ij}|| =
\begin{pmatrix}
c_{11} & c_{12} & 0 \\
c_{12} & c_{11} & 0 \\
0 & 0 & c_{66}
\end{pmatrix}.
\tag{4.9}
$$

For 2D, the Born criteria change accordingly:

$$c_{11} > |c_{12}|, \quad c_{66} > 0. \tag{4.10}$$

For 2D materials, the matrix of compliance coefficients s_{ij} has the form

$$||s_{ij}|| = \begin{pmatrix} s_{11} & s_{12} & 0 \\ s_{12} & s_{11} & 0 \\ 0 & 0 & s_{66} \end{pmatrix}. \tag{4.11}$$

Using Hooke's law and Eqs. (4.9) and (4.11), we have the following relationship between the components of the compliance coefficients and the stiffness coefficient for the 2D hexagonal structures:

$$s_{11} = \frac{c_{11}}{c_{11}^2 - c_{12}^2}, \quad s_{12} = -\frac{c_{12}}{c_{11}^2 - c_{12}^2}, \quad s_{66} = \frac{1}{c_{66}}. \tag{4.12}$$

For an orthorhombic crystal, the number of independent stiffness constants is nine: c_{11}, c_{12}, c_{13}, c_{22}, c_{23}, c_{33}, c_{44}, c_{55} and c_{66}. In the case of 2D materials for orthorhombic crystals, the matrix of stiffness coefficients and the matrix of compliance coefficients should be respectively written as

$$||c_{ij}|| = \begin{pmatrix} c_{11} & c_{12} & 0 \\ c_{12} & c_{22} & 0 \\ 0 & 0 & c_{66} \end{pmatrix} \tag{4.13}$$

and

$$||s_{ij}|| = \begin{pmatrix} s_{11} & s_{12} & 0 \\ s_{12} & s_{22} & 0 \\ 0 & 0 & s_{66} \end{pmatrix}. \tag{4.14}$$

And the stability criterion is written as

$$c_{11} > 0, \quad c_{22} > 0, \quad c_{11}c_{22} > c_{12}^2, \quad c_{66} > 0. \tag{4.15}$$

The relationship between the components of the compliance coefficients and the stiffness coefficient has the form

$$s_{11} = \frac{c_{22}}{c_{11}c_{22} - c_{12}^2}, \quad s_{22} = \frac{c_{11}}{c_{11}c_{22} - c_{12}^2},$$

$$s_{12} = -\frac{c_{12}}{c_{11}c_{22} - c_{12}^2}, \quad s_{66} = \frac{1}{c_{66}}. \tag{4.16}$$

For all considered syngonies, calculations of elasticity coefficients are carried out in a linear mode and only for stable structural configurations.

4.3 Tensile Strength

For the simulation of active deformation of the solid system under consideration, different simulation techniques can be applied. The main points is to follow the basic concepts of continuum mechanics such as energy conservation. The deformation scheme will affect the final results for problems of tension and fracture of the solid structures, and deformation scheme for each problem should be carefully chosen.

Here, several examples of the simulation of deformation and fracture will be considered:

(i)   deformation of the 2D structures;
(ii) deformation of the 2D structures taking into account the possibility of rippling;
(iii) deformation of graphene/metal composites.

Let us consider the graphene/metal composite composed of a graphene network (16,128 carbon atoms in total) and Ni nanoparticles inside the cavities of the graphene flakes (3,008 Ni atoms in total). In this example, the size of the simulation cell along x, y, and z directions are 100 Å, 101 Å, and 85 Å, respectively. A more detailed description of the starting configuration of the Ni/graphene composite is presented in [234, 236, 237].

LAMMPS simulation package is used for this work. The interactions between Ni and C atoms are referred to the Morse potential with the parameters $D_e = 0.433$ eV, $R_e = 2.316$ Å, $\beta = 3.244$ 1/Å [77, 121]. The Ni-Ni interactions are also fitted using the Morse potentials with parameters $D_e = 0.4205$ eV, $R_e = 2.78$ Å and $\beta = 1.4199$ 1/Å [88]. The C-C interactions are described with AIREBO potential.

Two different techniques can be used to simulate the deformation of the system: incremental and dynamic loading [306]. For the former, tension is followed by relaxation during which boundaries of the simulation cell are fixed (to relax stresses without tensile loading). Note that the external pressure is not decreased to zero, but remains at the same level as during deformation step. This scheme allows the redistribution of internal stresses in the structure, since the strain rate in MD simulation is too high. For dynamic loading, strain monotonically increases with time, and borders of the simulation cell are displaced continuously.

To compare the results of dynamic and incremental uniaxial tension, Fig. 4.12 shows the stress-strain curves obtained at 0 K. Two different strain rates are also presented to show the difference in the obtained stress-strain curves.

For the same strain rate $\dot{\varepsilon} = 5 \times 10^{-3}$ ps^{-1}, the tensile strength of the Ni/graphene composite under dynamic load is higher than under incremental load. However, critical stress is close in both cases. At a strain rate $\dot{\varepsilon} = 5 \times 10^{-4}$ ps^{-1}, a difference in tensile strength is not observed, but the formation of long carbon chains occurs faster under incremental load, which results in faster fracture. As mentioned in [306], under dynamic load covalent networks exhibit brittle fracture instead of ductile because of insufficient structure relaxation.

Temperature significantly affects the tensile strength of the Ni/graphene composite which can be seen in Fig. 4.13. At 300 K, the strength of the composite is 17-23% lower than at 0 K. The method of applying tension has no effect.

Young's modulus (E) of the obtained composite are 219 GPa (at 0 K and $\dot{\varepsilon} = 5 \times 10^{-3}$ ps^{-1}), 313 GPa (at 300 K and $\dot{\varepsilon} = 5 \times 10^{-3}$ ps^{-1}) and 218 GPa (at 0 K and $\dot{\varepsilon} = 5 \times 10^{-4}$ ps^{-1}).

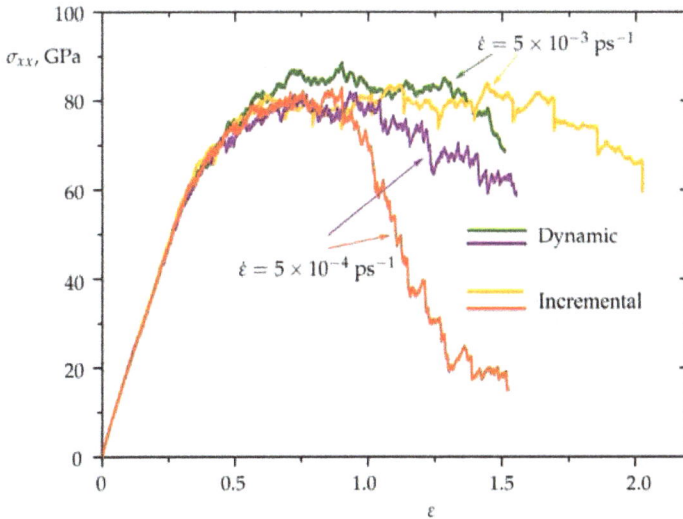

Fig. 4.12. Stress-strain curves during uniaxial dynamic (green and violet curves) and incremental (orange and red curves) tension along x-axis for strain rate $\dot{\varepsilon} = 5 \times 10^{-3}$ ps^{-1} and $\dot{\varepsilon} = 5 \times 10^{-4}$ ps^{-1} at 0 K. Reprinted with permission from [138].

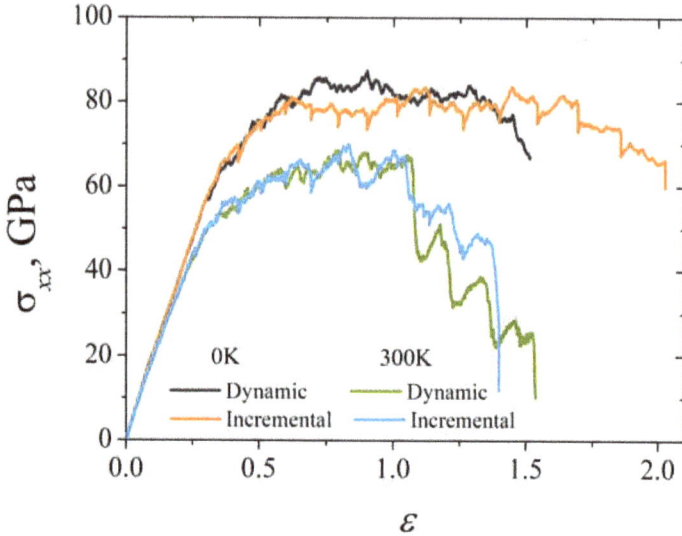

Fig. 4.13. Stress-strain curves during uniaxial tension for dynamic (black and green curves) and incremental (orange and blue curves) deformation at 0 and 300 K. Strain rate is $\dot{\varepsilon} = 5 \times 10^{-3}$ ps^{-1}. Reprinted with permission from [138].

4.4 Hardness

For the example of the calculation of hardness, Ni single crystal and Ni with composite coating of different thicknesses are considered. The geometrical dimension of the Ni single crystal is $L_x = L_z = 15.0$ nm and $L_y = 10.5$ nm. Figure 4.14 shows the schematic of the nanoindentation. PBCs are applied to the simulation box along the x and z directions. Along the y direction the cell size is increased to allow simulation of the nanoindentation process.

To achieve thermodynamic equilibrium in the system at 10 K, the simulation cell is exposed during two stages: (1) exposure is carried out using a Nose-Hoover thermostat in an isothermal-isochoric NVT ensemble, and (2) exposure with an isothermal-isobaric NPT ensemble. Each exposure stage is performed for 20 ps with a time step of 2 fs. During nanoindentation, the system is evolved in a microcanonical NVE ensemble using a Langevin thermostat.

A spherical indenter of two radii of 10 and 20 Å is located above the center of the composite and far enough that there is no interaction force between indenter tip and composite. The virtual indenter applies only repulsive forces to the sample and is considered a rigid body. The indenter is

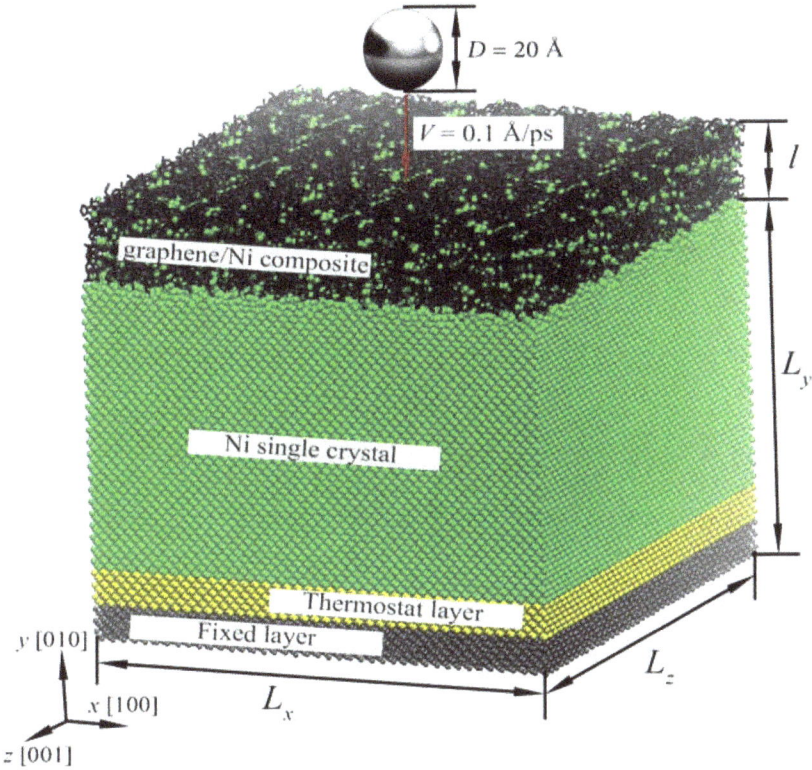

Fig. 4.14. Simulation scheme of the indentation process. Nickel atoms are shown in green, and carbon atoms in black. A layer of atoms (colored grey) at the bottom is fixed to avoid any transitional or rotational motion of the structure during indentation. The next layers of atoms (colored yellow) are set as thermostat layers. The initial position of the spherical indenter is located above the center of the composite and far enough that there is no interaction force between indenter tip and composite. Reprinted with permission from [136].

moved along the y-axis at a speed of 0.01 nm/ps. To avoid any transitional or rotational motion of the structure during indentation, a layer of atoms with a thickness of 1.0 nm at the bottom of the simulation cell (grey atoms in Fig. 4.14) is fixed along the y-axis. The next 1.0 nm thick layer (yellow atoms in Fig. 4.14) has a fixed temperature of 10 K controlled by the Langevin thermostat.

Nanohardness is determined using the Oliver-Farr method [193]. Hardness (H) and reduced Young's modulus (E^*) are calculated from the load/unload curves recorded during indentation. The equations for the corresponding calculations are given below:

$$H = \frac{F}{A_c}, \tag{4.17}$$

$$E^* = \frac{\sqrt{\pi}}{2\beta} \frac{S}{\sqrt{A_c}}, \tag{4.18}$$

where F is the indentation force, A_c is the contact area between the indenter and the sample surface, $\beta = 1$ for the spherical indenter [44], and S represents the contact stiffness ($S = dP/dh$) which is determined by the slope of the unloading curve [193]. The contact area (A_c) is calculated using [59, 162]:

$$A_c = \pi(2Rh_c - h_c^2), \tag{4.19}$$

where R is the indenter radius and h_c is the contact depth.

The Oliver-Farr method is most commonly used to determine the micro- and nanohardness of nanocoatings, where the accuracy of the measurements is important. In order to exclude the substrate effect on the obtained results, the maximum depth of the nanoindenter cannot increase beyond 0.1 of the total thickness of the sample.

Firstly, it is very important to analyze the effect of indenter radius on the obtained results. Figure 4.15 shows the indentation force F and hardness H as the function of the indentation depth h for three indenter radii (10, 20 and 30 Å) during loading and unloading. The critical force for dislocation nucleation increases with increasing indenter radius (see Fig. 4.15a),

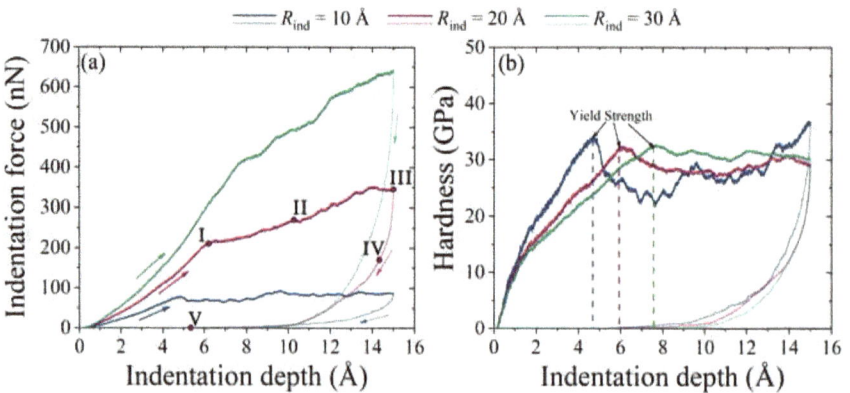

Fig. 4.15. (a) Indentation force F and (b) hardness H as a function of the indentation depth h for Ni single crystal. Reprinted with permission from [136].

while hardness decreases with increasing indenter size, which corresponds to previous results [161, 231]. The $H - h$ curves for R_{ind} equal to 20 and 30 Å correlate well and a further radius of 20 Å can be taken. Figure 4.15b shows that pop-in events are more frequent on the hardness curve for a smaller indenter radius, which increases the measurement error, while measurements with a larger indenter radius give a more accurate result, but calculations are longer. For $R_{ind} = 20 - 30$ Å, the average hardness is 29-30 $\pm$ 2 GPa, and for $R_{ind} = 10$ Å the average hardness is 36 ± 5 GPa. The yield strength of Ni is almost independent on the indenter radius and equal to 32–33.5 GPa. However, the first defects appear faster for the indenter radius of 10 Å at indentation depth of 4.7 Å. For a 20–30 Å radius, the plastic yield stage begins at an indentation depth of 5.8 and 7.5 Å, respectively (see Fig. 4.15b).

As the indenter made contact with the substrate, the indentation force increased smoothly until the material deformed plastically. The pop-in (Yield Strength point in Fig. 4.15b) events represent the activation and motion of dislocations. The first dislocations are generated homogeneously under the indenter, which is accompanied by a decrease in force and hardness. The yield strength in Fig. 4.15b corresponds to the nucleation of the first defects, after which the plastic yielding begins. During the indentation, interaction between defects took place, which results in strengthening and increase in the indentation force and hardness. All the results are presented for the indenter radius of 20 Å.

4.5 Structure Characterization

Visualization of the obtained results plays an important role in the process of MD simulations. During the simulation, a program is working with a giant amount of numbers – atomic coordinates, stresses, energies, velocities, to name a few. But for humans, these numbers do not provide any information about the system as a whole. We cannot imagine what happens in the system during the simulation just from these numbers. Therefore, special methods are developed to analyze the results of the simulation. For crystalline solids, which have the greatest variety of possible structures and defects, these methods are of particular importance. If there is a system (a crystal structure) consisting of one or more types of atoms with the given coordinates, then it can be displayed on the screen, depicting the atoms as certain geometric bodies (usually in the form of spheres) in accordance with their relative positions. In this case, each type of atom can be assigned a certain color, the connections between them can be depicted, etc.

The simplest method of structure analysis is the visualization of all the atoms of the system at a given moment in time. It allows us to visually determine the structure, the shape of the sample, lattice defects, etc. The principle of visualization is simple: an image of atoms is created on the screen in accordance with their coordinates. Atoms can be represented as spheres of different radii or colors.

4.5.1 *VMD Visualization Tool*

The VMD visualization tool uses a file, containing data on the material structure. The program works in a three-window mode, as shown in Fig. 4.16, where you can see a text, graphic and menu window. The graphic window visualizes the structure, and in the text window you can enter commands that control the visualization, but similar actions can be performed through the menu window, so below we will describe the visualization of structures through the menu bar. The main menu has seven commands: 'File', 'Molecule', 'Graphics', 'Display', 'Mouse', 'Extensions', 'Help'. The program works with files with different extensions, such as '.XYZ', '.PDB', '.lammpstrj', to name a few.

Graphene, graphite and carbon nanotubes (CNTs) can be created using VMD (see Fig. 4.17). To create a carbon structure, select the menu 'Extensions' $\longrightarrow$ 'Modeling' $\longrightarrow$ 'Nanotube Builder'. The window shown in Fig. 4.17 allows you to choose graphene or CNTs, atom type, and lattice parameter. The 'Topology Building Options' section allows you to set the bond type (C–C) and the lattice parameter (in this example, 0.1418 nm for graphene).

In the 'Nanotube Building Options' field, you can create a carbon nanotube of different lengths and chiralities. Each nanotube has two indices, m and n, which determine the chirality of CNT: for an armchair nanotube, $m = n$, for a zigzag nanotube, $m = 0, n \neq 0$, and for a chiral nanotube, both indices should be specified. The user can specify two chirality indices, 'Nanotube chiral index n' and 'Nanotube chiral index m'. In addition, you can specify the nanotube length in nanometers. After the parameters are specified, the 'Generate Nanotube' button allows the user to obtain a CNT for visualization and further work with the nanotube.

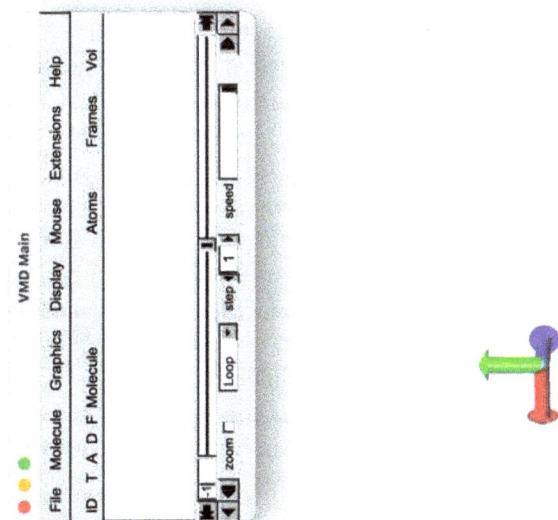

Fig. 4.16. Three-window mode of VMD: text, graphic and menu window.

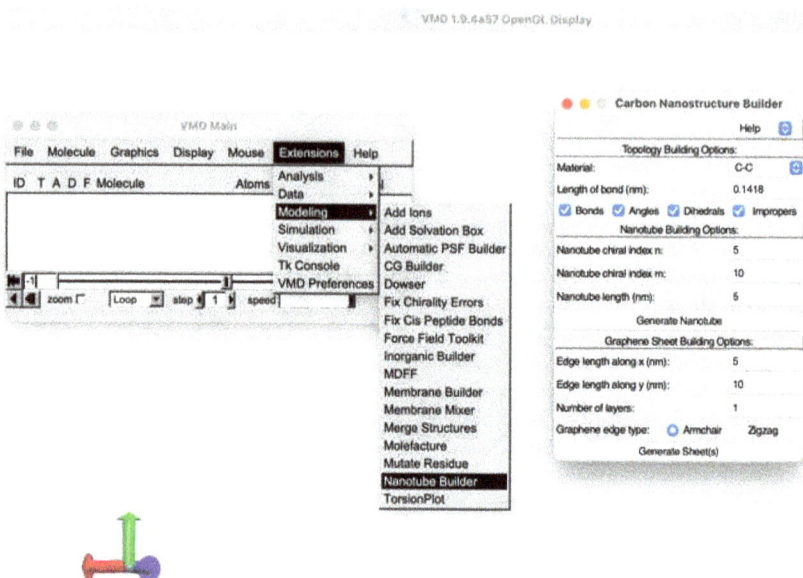

Fig. 4.17. 'Nanotube Builder' window.

In the 'Graphene Sheet Building Options' section, you can build a graphene sheet or a graphite 3D crystal. The first two fields specify the size of the graphene sheet in nanometers, and the last field specifies the number of layers ('Number of layers'): accordingly, if the layer is 1, then it is graphene, and if you specify a larger number of layers, then we can get multi-layered graphene or even graphite. In addition, you can choose which edge of graphene (zigzag or armchair) will be oriented along the x-axis, and which along the y-axis. After the parameters are set, the 'Generate Sheet(s)' button allows you to create graphene for visualization and further work with the structure.

Figure 4.18 presents CNTs of different chirality obtained with VMD as an example. Parameters n and m affect the diameter of the CNT (see Chapter 2). As can be seen, CNTs can be visualized in a different way. For example in Fig. 4.18a, visualization was changed through the 'Graphics' $\longrightarrow$ 'Representations': 'Drawing method' – 'Dynamic Bonds'. In Fig. 4.18b,c, CNTs are presented as two 'Drawing method': 'Dynamic Bonds' for covalent bonds and VDW for atoms representation. Also, different colors can be used

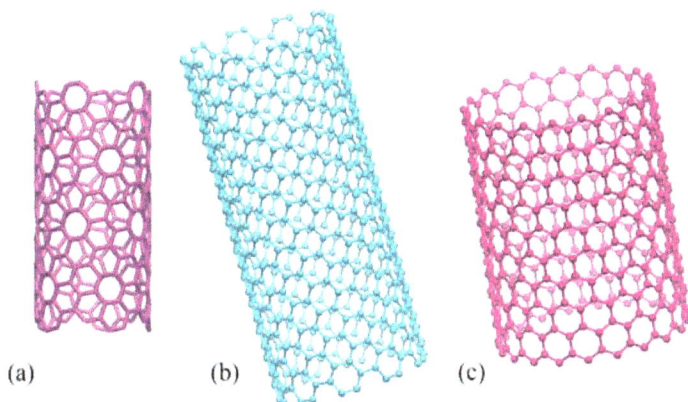

Fig. 4.18. CNTs obtained with VMD: (a) chiral with $n = 5$, $m = 10$, and length of 2 nm; (b) armchair with $n = 10$, $m = 10$, and length of 3 nm; (c) zigzag with $n = 0$, $m = 20$, and length of 2 nm.

for visualization. Detailed methods for visualization can be found in the VMD website and tutorials (https://www.ks.uiuc.edu/Research/vmd/).

In addition to visualization, VMD allows us to analyze the structure of the material, for example, the lattice parameter or valence angle for carbon nanostructures. This can help to analyze the deformation mechanisms for nanostructures, when changes of the interatomic bonds and valence angles are highly important.

Another important feature of VMD for structural analysis is the calculation of the radial distribution function, which can help to characterize the atomic ordering in the structure. The radial distribution function characterizes the correlation in the arrangement of particles of a gas, liquid or solid. A random arrangement of atoms is possible only in gases, when the interaction forces between atoms can be ignored. In liquids and solids, atoms cannot be at an arbitrary distance from each other, since their packing is sufficiently dense. The probability of finding an atom at a point in volume V depends on the point at which another atom is located. Such a probable connection between the mutual arrangement of atoms (their correlation) is quantitatively described by the function $g(\vec{R}_1, \vec{R}_2)$, where $\vec{R}_1$, $\vec{R}_2$ are radius-vectors characterizing two elemental volumes near the atom.

The radial distribution function determines the probability density of detecting any atom at a distance R from the selected atom. Since repulsive

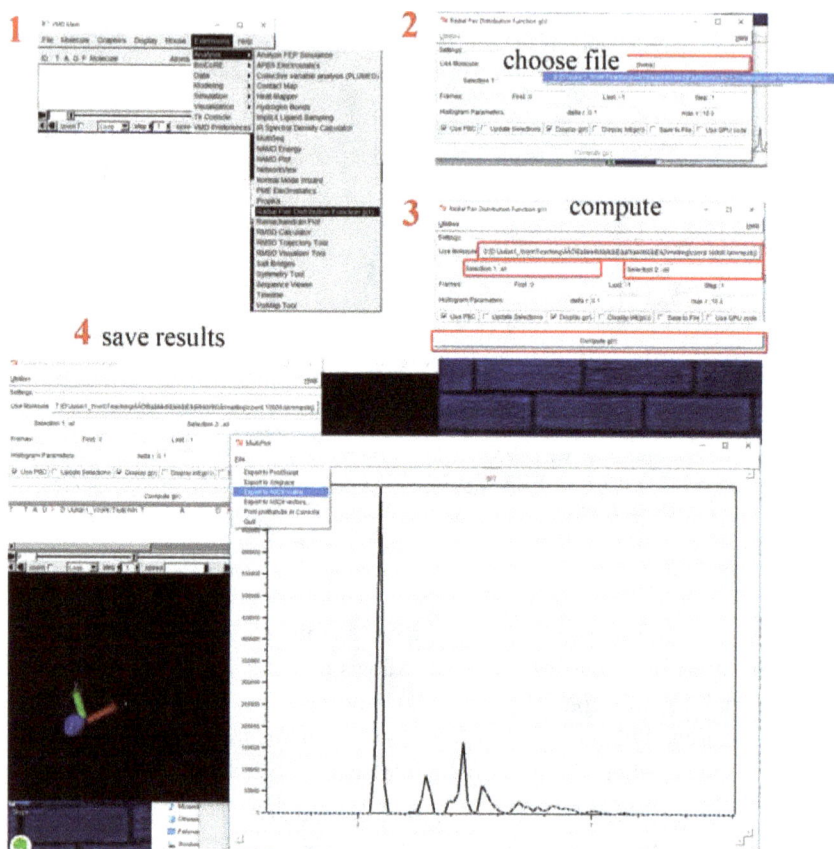

Fig. 4.19.   Calculation of the radial distribution function in VMD.

forces prevent mutual penetration of atoms, the radial distribution function is zero in the interval $0 \leq 0 < 2r$, where r is the radius of the atom. When $R \to \infty$, the radial distribution function tends to unity. This means that at large distances, the positions of the atoms become uncorrelated.

Figure 4.19 presents the process of the calculation of the radial distribution function in VMD.

4.5.2 *OVITO Visualization Tool*

As with VMD, OVITO visualization tool uses a file containing data on the material structure. The main window is presented in Fig. 4.20. This

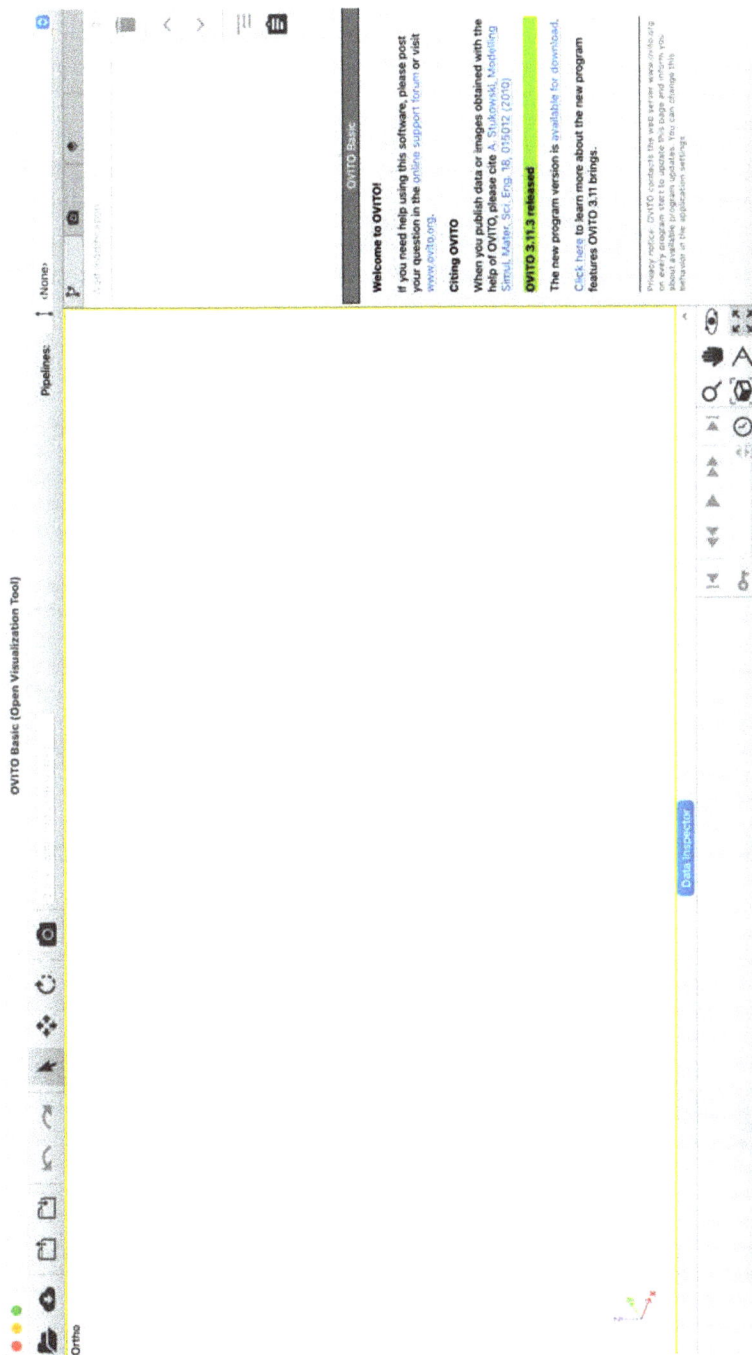

Fig. 4.20.   Window mode of OVITO.

visualization tool allows us to present very different characteristics of the structure: common-neighbor analysis; stress or energy distribution; dislocation analysis, to name a few.

Figures 4.21 and 4.22 present the per atom coloring of the structure in accordance with the potential energy per atom calculated during tensile deformation of the metallic nanofiber. To show the energy distribution, energy per atom should be calculated with LAMMPS command and written to the file. Then, in OVITO with the menu "Color coding" this energy

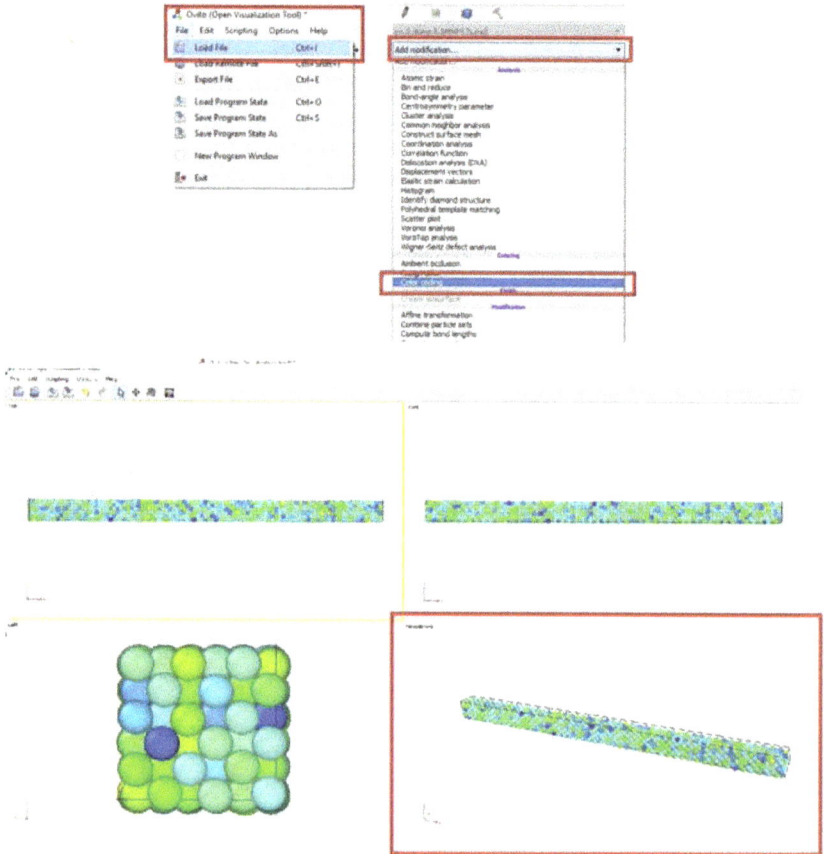

Fig. 4.21.   Per atom coloring of the structure in accordance with the potential energy per atom.

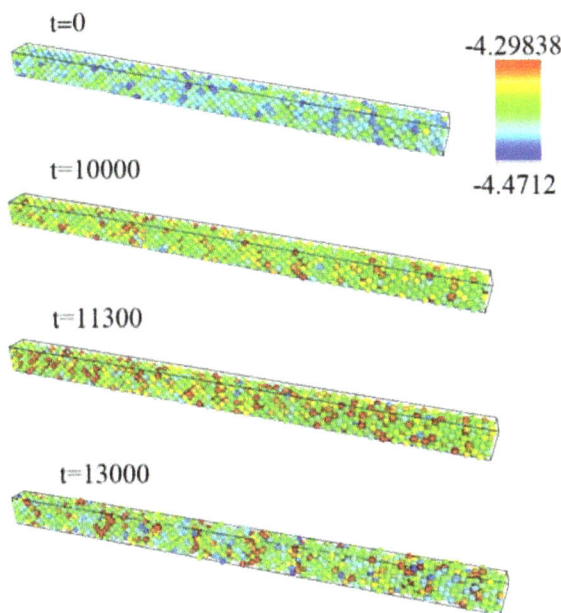

Fig. 4.22.   Per atom coloring of the structure in accordance with the potential energy per atom for metallic nanofiber under tension.

per atom can be easily visualized. Atoms with an energy (or stress) value included in each interval are represented by their own color during visualization. Then, these atoms are visualized as large-diameter circles. As a result, the sample will be in different colors in accordance with the energy values, like Fig. 4.22.

What can be found from Fig. 4.22? We can see that during tension overall energy of the sample was increased; however, there is no critical increase of the energy, just some atoms have high energy (red) which means that, for example, there was no fracture of the nanofiber yet. Analysis of the energy distribution is very important for investigations of defects, grain boundaries, melting of the structure, etc.

Figure 4.23 presents the stress distribution per atom during the uniaxial tensile deformation of two different graphene networks (a) and (b). The red atoms are the straightest graphene flakes, while the straightest carbon chains are seen at high strain levels. This allows us to understand how the

(a)

(b)

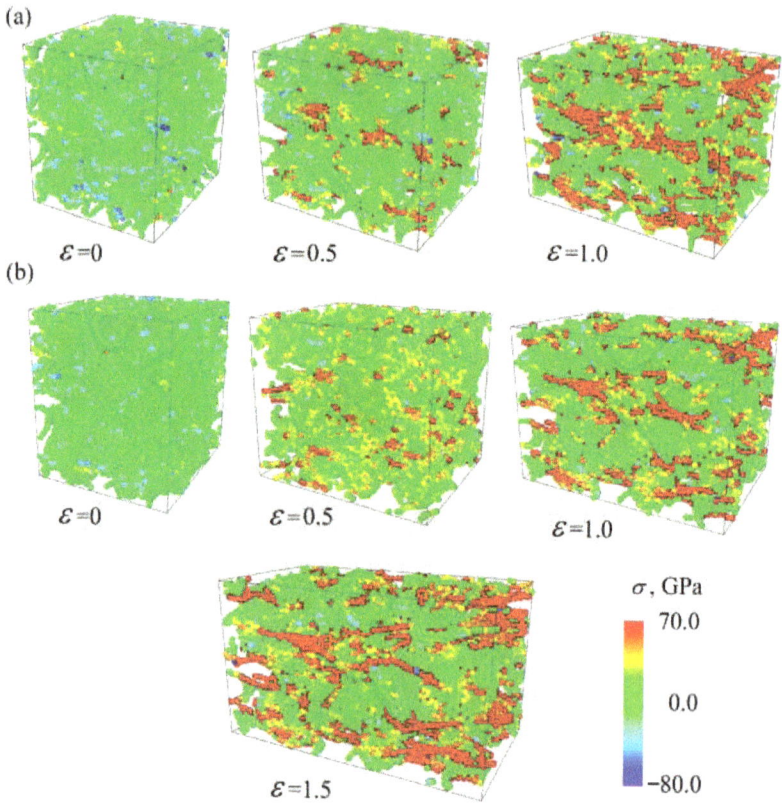

Fig. 4.23. Stress distribution per atom during the uniaxial tensile deformation of two different graphene networks. Reprinted with permission from [127].

deformation of graphene network occurred, to find the most stressed places and understand the weakest places in the structure.

OVITO also allows us to analyze the dislocation dynamics in metal structures. Figure 4.24 presents the snapshots of (a) Ni single crystal and (b-d) Ni with a composite graphene coating of different thicknesses at an indentation depth of 0.0, 6.1, 8.0 and 15.0 Å. For coating thickness 1 nm, the first dislocations are generated uniformly at the interface at the indentation depth of $h = 5.8$ Å, close to that of pure Ni. For coating thickness of 5.1 nm, dislocations are not generated even at high indentation force. Graphene

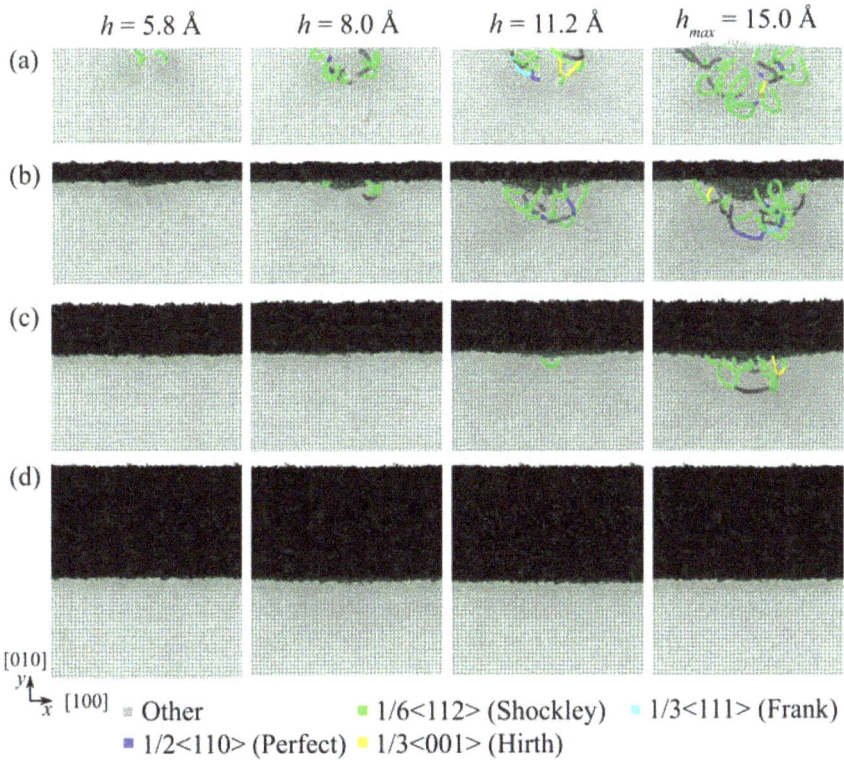

Fig. 4.24. Snapshots of the atomic structure at different indentation depths for Ni single crystal (a) and Ni with a composite graphene nanocoating with the thickness of 1 (b), 2.4 (c), and 5.1 nm (d). Reprinted with permission from [136].

network resists large indentation stresses and redistributes them throughout the entire structure of the composite coating. In this case OVITO allows us to show the dislocation generation under different coating thickness and understand the strengthening mechanisms.

Chapter 5

Future Applications

5.1 Electronics and Sensors

Currently, for the development of new powerful electronic devices, the size of their main components should be reduced to nanoscale, thereby increasing their productivity. The future in the development of nanoscale electronics is associated with carbon-based materials. In particular, fullerenes, graphene, carbon nanotubes (CNTs) and their derivatives have already found their application in the manufacture of field-effect transistors, solar batteries and organic light-emitting diodes.

Flash memory traditionally occupies one of the leading positions among various types of non-volatile memory elements. Recently, it has been shown that graphene, multi-layered graphene, and graphene oxide are promising replacements for the material used for the floating electrode in flash memory elements. Graphene is a semi-metal and therefore can capture a significantly greater charge than polysilicon, silicon nitride, and semiconductor nanocrystals. Due to the large work function of carriers, a high potential barrier is formed between graphene and silicon, which provides a significant increase in the storage time of information on the graphene electrode. It is also important to use graphene to create flexible memory devices. The operation of flash memory elements is based on the capture of carriers on the floating electrode in a planar field-effect transistor. The low effective mass for carriers in multi-layered graphene and zero effective mass for carriers in graphene provides picosecond relaxation time for nonequilibrium carriers and their movement through the potential barrier between graphene layers, which contributes to the speed of operation of graphene-based devices.

Fullerenes can be used in areas such as nanoelectronics, sensors, additives for lubricating oils, etc. CNTs are used in transistors, light-emitting diodes, etc. To note, the use of CNTs for nanoelectronic devices is related not only to their electrical, but also to their thermal properties, since the performance and stability of nanoscale electronic devices largely depends on the efficiency of their cooling. One of the most promising areas is the use of single-walled CNTs as a nanoreactor for the synthesis of fundamentally new 1D structures inside nanotubes. The 1D structures synthesized in this way may have their own interesting properties. The possibility of using a hybrid nanomaterial (1D structure inside single-walled CNTs), in which the desired properties are achieved due to the synergy of properties, is attractive. Due to their own unique properties, single-walled CNTs are ideal nanoreactors.

To date, there are many literature reviews on graphene application [3, 30, 89, 113, 171, 176, 191, 276, 282, 286], thus, in the present chapter only main thoughts on this are presented. The short overview of various applications of CNTs and graphene is presented, while considerable attention is paid to graphene application in hydrogen storage. Figure 5.1 shows the areas of graphene application.

Despite the absence of a band gap in graphene resulting in certain difficulties for its wide application in transistor technologies, its unique properties (high carrier mobility and low electrical resistance, as well as extremely small thickness and stability) allow it to be successfully used in electronics. The key performance indicator for transistors is the cutoff frequency. The first such devices were created on the basis of graphene synthesized by thermal decomposition of silicon carbide.

There are two candidates to be new non-silicon transistors – graphene and CNTs. As in many other nanodevices, the purity and quality of the original carbon structures have a great influence on the characteristics of the device. The existing methods of their production do not yet allow full control over the state of the edges, the presence of defects and numerous impurities; however, the possibilities of creating transistors based on graphene, graphene nanoribbons and CNTs are being actively explored all over the world. First of all, the production of new transistors should affect the development of computer engineering and overcome Moore's law, according to which in the near future (by 2020) even all existing nanotechnologies and developments will not allow any increase of computer performance.

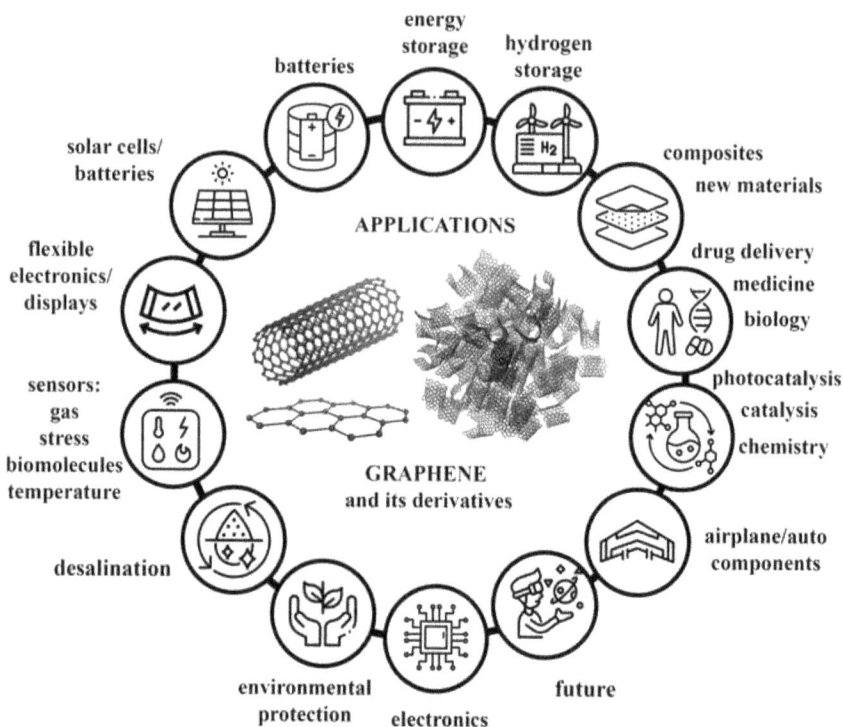

Fig. 5.1. Applications of graphene.

The use of CNTs in the development of computer technologies plays an important role in the electronics industry. CNTs can be used for production of fairly flat displays, compact computer equipment, etc. Due to the fact that the electronic properties of nanotubes change depending on chirality, it is possible to create semiconductor heterostructures based on CNTs. In such a structure, a defect appears that leads to a change in the chirality of the CNTs and, consequently, to a change in the type of conductivity of the semiconductor/metal. To date, CNTs act as elements of vacuum devices as sources of electrons, are used as a basis in the designs of the memory architecture of nanocomputers, are used as field-effect transistors, etc.

Graphene is characterized by high mobility of electrons and holes at room temperature. On the other hand, one of the disadvantages of graphene is that graphene transistors can switch between closed and open states very quickly. In this case, a device containing billions of graphene transistors will be characterized by great energy losses. This can be overcome by using nanoribbons instead of graphene, which will lead to an increase in the ratio of the conductivity of the closed and open states of transistors. Unlike conventional transistors made of semiconductors with a certain band gap, a graphene transistor always has residual conductivity. This is a critical factor in determining the role of graphene in modern electronics, so much effort is being put into creating a band gap in the graphene structure.

Graphene-based transistors can operate not only in electric fields, but also in magnetic fields: in such devices, electricity begins to flow faster under the influence of a magnetic field. Accordingly, by changing the strength and direction of the magnetic field, it is possible to control the current through graphene. Bilayer graphene can also be used for new transistors: their energy zones are such that the density of electrons that can be placed near the edges tends to infinity, which means that with the application of a small voltage to the transistor, the electrons will begin to tunnel. Therefore, in bilayer graphene, it is possible to obtain the currents necessary for the operation of electronics with low energy consumption.

One of the popular applications for CNTs and graphene could be flexible electronics. Currently, fully flexible circuit board technologies are based on conductive polymers, but the mobility of charges, and therefore the speed of response, is much lower than that of silicon and other semiconductors used in solid-state electronics. Such flexible transistors are needed to meet modern requirements for the creation of personal electronic devices – small size combined with high functionality, flexibility of sensor devices combined with high strength. New technologies already make it possible to make electrical circuits on large-sized graphene films to create stretchable transparent electrodes. Future applications of this technology include computers in the form of clothing, flexible transparent displays, touch panels, folding electronic paper and transformable electronics.

A new type of graphene-based device is a supercapacitor with an energy density ten times higher than traditional capacitors and a pulse discharge power up to ten times higher than that of batteries. A supercapacitor is a molecular energy storage device and is of great interest for accumulating and storing energy, having high power, a fast charging process and unique cyclic stability. Supercapacitors provide good compatibility with lithium-ion batteries in devices that require both high energy (lithium-ion batteries) and high power (supercapacitors), thereby increasing the breakdown voltage, expanding the efficiency of using electrical energy and increasing the battery life. According to the classification, supercapacitors are divided into two main types – electrical double-layer capacitors and pseudo-capacitors. In electrochemical double-layer supercapacitors, simple electrostatic attraction is realized between ions accumulated on the electrode/electrolyte surface, with activated carbon chosen as the electrode.

The most important advantages of supercapacitors in comparison with lithium-ion and lithium-polymer batteries are high charging speed, efficiency and huge resource. Supercapacitors are capable of storing a large amount of energy in a short period of time, which allows users to reduce the recharging time to a minimum.

Carbon-based materials are widely used as electrode materials in double-layer capacitors due to their unique physicochemical properties. In pseudo-capacitors, electrons are additionally involved in fast Faradaic interactions and move to or from the valence bands of the redox cathode or anode reagent. Various transition metal oxides and conductive polymers are used as electrodes in pseudo-capacitors; however, graphene and other carbon materials have also great potential to produce supercapacitors with improved characteristics. The key point in the use of carbon materials is to increase the temperature limit of graphene and find a new method to combine the carbon matrix with a doping agent to further enhance and stabilize superconductivity.

The use of graphene for sensors is explained by its large specific surface area, unique optical properties, high mobility of charge carriers, and exceptional electrical and thermal properties compared to other carbon allotropes. Another advantage of graphene sensors is their ability to be customized according to the requirements. For example, in strain sensors, properties such as the detection limit, maximum sensitivity range, signal response, and response reproducibility play a key role. In electrochemical sensors, the large specific surface area of graphene facilitates the capture of the biomolecule. Another advantage of graphene is its low environmental impact and harmlessness, which makes it more popular for sensing than other materials. Analysis of the results shows that most graphene-based sensors are able to achieve high sensitivity in a wider range.

Graphene is capable of sensing the adsorption of even one molecule. The attached gas molecules, depending on their charge and the type of conductivity of the graphene film, behave as donors and acceptors, i.e., they change the concentration of mobile charge carriers. As a result, depending on the type of adsorbed molecule, a decrease or increase in the film resistance can be observed. It should be noted, however, that one of the serious drawbacks of the graphene gas sensor is the lack of selectivity. Indeed, from the change in conductivity it is impossible to tell which molecule was "absorbed" by the graphene surface. Moreover, some molecules make contributions of the opposite sign, so the total change in resistance can be close to zero.

Although strain sensors with different materials have been invented and developed for quite some time, graphene-based sensors have proven to be an excellent alternative to existing devices. The advantage of using graphene over other conductive materials to detect strain is the generation of a pseudo-magnetic field, which makes it possible to detect the change in the electronic structure during deformation. The manometric coefficient is an important parameter that is used to determine the efficiency of the manufactured strain gauges. As a result, the electrical shift is determined based on the mechanical deformations. This is due to the fact that since graphene-based sensors include different materials combined to make them, the change in resistivity per unit length change caused by deformation determines the amount of deformation that must be taken into account for the material.

The development of fuel cell devices and technology requires the new hydrogen sensors to ensure their safe operation. CNTs are an ideal material for sensor components due to their inherent durability and good electronic properties. One of the first hydrogen sensors with CNTs was created in 2001 by adding Pd nanoparticles to single-walled CNTs. This sensor showed significant changes in conductivity when adding a small amount of H_2 and could operate at room temperature [179]. After the sensor was placed in a hydrogen environment, an increase in the conductivity of the CNTs was observed. In a hydrogen environment, Pd reacts with H_2 and becomes palladium hydrate. Dissolved hydrogen combines with atmospheric oxygen and leads to the formation of H_2O, restoring the electrical characteristics of the sensor. The use of CNTs in chemical sensors is limited due to the difficulty of integrating 1D structures into electronic devices.

Graphene is a promising material for detecting molecules of various gases, as well as biological substances. Charge transfer between graphene and adsorbed molecules is caused by chemical interaction between them, as a result of which the Fermi level, carrier density and resistance change. It is believed that the sensor properties of films depend on the presence of various functional groups on their surface. It is also known that the efficiency of sensors can be increased by increasing their surface area.

5.2 Hydrogen Storage

One of the main obstacles to the use of hydrogen as a universal and environmentally friendly fuel is the lack of effective methods for its accumulation. Among the well-known methods for hydrogen storage are adsorption at low temperatures, under high pressure, in a liquid state, in the form of metal hydrides and intermetallic compounds. Currently, none of the existing methods of hydrogen storage allows an increase in either the gravimetric capacity of hydrogen storage or the volumetric density of the hydrogen. To date, it has not been possible to solve the problem of reversible physical sorption of hydrogen at room temperature in such a way that the stored amount of hydrogen would be sufficient for its effective practical usage. Therefore, over the past decades, a search has been carried out for materials that can be used as cells for hydrogen transportation and storage. Most often, hydride-forming metals, intermetallic compounds and activated carbon prepared in various ways are considered as sorbents. However, in recent years, fullerenes, carbon nanofibers and nanotubes, and different graphene derivatives are considered promising for use as hydrogen-storage devices.

In total, three main aspects need to be taken into account when developing effective hydrogen carriers: (a) storage capacity, (b) stability and safety during hydrogen storage, and (c) hydrogenation/dehydrogenation kinetics. Since graphene has a low weight and can form porous structures, it is possible to increase the gravimetric hydrogenation capacity of the material. In addition, it has been shown that van der Waals forces in bulk carbon increases the safety parameters of hydrogen storage, and at the same time the gravimetric capacity of its accumulation, which allows carbon nanostructures to meet the basic requirements for hydrogen carriers. Since graphene is a flexible material and can be transformed into 3D configurations where the structural elements are linked by van der Waals forces, this material is considered promising for hydrogen storage and transportation. Moreover, the properties of graphene-based structures can be effectively controlled by external influences, such as deformation, which allow further improvement of the sorption properties of the material.

The maximum gravimetric density can be obtained for graphene with a chemical absorption rate of 8.3%, which corresponds to the formation of a fully hydrogenated graphene with a stoichiometry of 1:1 for C:H, which is called graphane. The hydrogenation of graphene was discussed in Chapter 2.

For physical adsorption, the volumetric density depends on the ability to compactly fill the graphene structure. To determine the hydrogen storage potential, it is promising to create 3D multi-layer graphene. Recently, a new 3D material has been developed composed of graphene layers connected by CNTs, acting as linking columns that stabilize the structure and hold the graphene layers at a given distance. This structure is called columnar graphene. From theoretical estimates, it has been shown that the energy of physical absorption is almost doubled compared to single-layer graphene and reaches a value of about 0.1 eV. In this case, the mass density increases to about 30-40% of the density of the single-layer structure, potentially reaching 8% at high pressure and low temperature, but remains within 3-4% at room temperature and high pressure. A mass density of 3% can be achieved at room temperature only under high pressure, while much higher values can be obtained at low temperature. It was experimentally shown that such layered structures can be created using graphene oxide.

The advantages of carbon carriers for hydrogen storage are large specific surface area, low density compared to intermetallics, chemical inertness, and resistance to cooling. To date, a variety of carbon carriers for reversible hydrogen sorption are known. Various systems, both traditional (activated carbon and graphite) and new (porous carbon threads, nanocarbon fibers and tubes), are considered as potential hydrogen sorbents. A distinctive feature of such materials is the fast kinetics and complete reversibility of hydrogen adsorption. The amount of adsorbed hydrogen is proportional to the specific surface area of nanostructured carbon. Therefore, with the maximum specific surface area of 1315 m^2/g, the maximum measured capacity of graphene for hydrogen adsorption is 2%(mass.). Crumpled graphene and other bulk carbon structures appear to be very promising materials for future hydrogen energy, since the pores can serve as hydrogen accumulation sites in the structure. Table 5.1 presents data on the gravimetric capacity of some carbon structures.

Table 5.1.   Gravimetric capacity of carbon polymorphs.

Material	Experimental data, %(mass.)	Theoretical data, %(mass.)
single-walled CNTs	4 (at 0.1 MPa and 77 K)	11.2 (at 10 MPa and 77 K)
	<1 (at 0.1 MPa and 295 K)	1–2 (at 60 MPa and 150–513 K)
muti-walled CNTs	5 (at 10 MPa and 300 K)	4.29–5.78 (at 10 MPa and 298 K)
CNT fiber	6.5–10 (at 12 MPa and 300 K)	–
graphene	0.9 (at 10 MPa and 298 K)	7 (at 1 MPa and 77 K)
		1.5 (at 30 MPa and 293 K)

5.2.1   *Fullerenes and Carbon Nanotubes*

The problem of fullerene hydrogenation is of fundamental importance in fullerene chemistry; however, fullerene is one of the most promising structures for storing hydrogen (along with graphene and CNT). Synthesis of fullerene hydrides has already been carried out in various ways. For example, the compound $C_{60}H_{36}$ with a small admixture of $C_{60}H_{18}$ was obtained as a result of the Birch reduction reaction. The same compound was obtained by means of the reaction of transferring a hydrogen atom between C_{60} and a molecule of 9,10-dihydroanthrocene. The compounds $C_{60}H_{36}$ and $C_{70}H_{36}$ were obtained from fullerenes using iodoethane heated to a temperature of 673 K as a hydrogen source, as well as by treating cold fullerenes with hydrogen at high pressure. As follows from the results of the experiments, the C_{60} molecule, under certain conditions, is capable of absorbing hydrogen atoms like a sponge, reversibly absorbing up to 17 atoms at a time. Apparently, the most direct method of hydrogenating fullerenes is associated with carrying out the reaction in the solid phase at high hydrogen pressure (up to 850 atm.) and elevated temperatures (about 600 K).

Fullerenes with alkaline elements can be effective for storing significant amounts of hydrogen. Functionalization of fullerenes leads to an improvement in a number of their properties. For example, adding C_{60} to $LiBH_4$ reduces the decomposition temperature of $LiBH_4$ and reduces the temperature and pressure requirements for its reduction.

Direct hydrogenation of fullerenes to obtain fulleranes (by analogy with graphene and graphane) is a well-known process. While a fullerene mixture typically retains an average of 31 hydrogen atoms per C_{60} molecule, fullerenes with hydrogen content greater than $C_{60}H_{36}$ have been obtained. For example, lithium-doped fullerenes can increase the reversibility and reduce the temperature and pressure required to obtain $C_{60}H_{48}$ with an average hydrogen content of 40 atoms per C_{60}. Hydrogen susceptibility can also be influenced by chemisorption by doping fullerenes with sodium.

Another way to increase hydrogen accumulation is to deposit hydrogen on fullerenes doped with alkali elements. The improvement of the hydrogenation process during doping is explained by the transfer of electron density charges to the fullerene from lithium atoms. It has been shown that the difference in charges polarizes the molecular orbital of hydrogen, which leads to electrostatic attraction between hydrogen and lithium located on the surface of the fullerene.

The process of hydrogenation of the fullerene molecule C_{60} is possible due to the fact that the valence electrons in each pentatomic molecule are delocalized, and their behavior is subject to the intramolecular laws of the pentatomic molecule. Experimental studies show that the destruction of the fullerite lattice begins already at a hydrogen content exceeding 36 atoms per fullerene molecule [246]. At a hydrogen concentration of more than 40 atoms, the peaks responsible for the fullerite crystal lattice are blurred in the X-ray diffraction patterns. This means that during the last stage of hydrogenation, the process of sequential rupture of the polymerizing bonds does not always occur. It is possible to rupture both polymerizing bonds at once, which leads to a violation of the long-range order in the fullerite lattice and the destruction of the lattice already at a hydrogen concentration of more than 36 atoms per fullerene molecule.

During complete hydrogenation of the C_{60} fullerene molecule, three hydrogen atoms enter each intermolecular hexatomic void. This leads to an increase in the diameter of the fullerene, complete saturation of the valence bonds and the destruction of the fullerite. When all hexatomic voids are filled with hydrogen atoms, the diameter of the C_{60} fullerene molecule can increase by 0.6, which is 8.6% of the diameter of the fullerene molecule.

Many experiments indicate that CNTs can be an effective medium for hydrogen storage. The possibility of storing hydrogen in single-walled nanotubes at room temperature has been demonstrated. An example is nanotubes synthesized in an arc discharge using hydrogen as a medium (instead of helium or nitrogen), and Ni, Co, Fe, and FeS as catalysts. The content of single-walled CNTs in the resulting material is estimated at 50–60% (the rest was mainly catalyst particles) [289]. Multi-walled CNTs synthesized by chemical vapor deposition at temperatures of 1050–1150°C using benzene as a hydrogen-containing substance and ferrocene as a catalyst also turned out to be a good medium for hydrogen storage. The nanotubes were filled with hydrogen for 12 hours at a hydrogen pressure of 150 atm. As a result, the amount of adsorbed hydrogen reached 6.5 wt.%, which corresponds to a specific capacity of the sample of 31.6 kg/m^3 [43]. The experiment showed that thorough cleaning of the samples led to a threefold increase in the amount of absorbed hydrogen. It is believed that the possible role of cleaning the samples is reduced to opening the heads of the nanotubes, facilitating more effective penetration of hydrogen into their internal cavities.

5.2.2   Graphene

Figure 5.2 presents three typical sites, H1 (hollow), B3 (bridge) and T1 (top), of molecular hydrogen sorption on graphene, which was found during exposure at 77 K. The lowest binding energy between graphene and H is observed when the distance between C and H is in the range from 2.9 to 3.2 Å. Hydrogen molecules are not rigidly bounded with the carbon structure and easily migrate over the surface of a graphene flake due to thermal fluctuation, and its configuration changes from one to another.

To understand the dynamics of the hydrogen sorption/desorption, snapshots of the structural units altogether with their potential energy are presented in Figure 5.3. The choice of hydrogenation temperatures is explained by best hydrogen sorption by carbon structures at 77 K (more than 2–3 wt.%), and secondly, with the requirement to achieve the optimal sorption capacity (more than 3 wt.%) at room temperature. At 77 K, hydrogen atoms are easily converted into hydrogen molecules H_2 and attach to a graphene flake by physical or chemical sorption. The hydrogen adsorption capacities depend considerably on the initial size of the hydrogen cluster. The specific surface area of the graphene flake should be big enough for a chosen number of H atoms and H_2 molecules. Here, the specific surface area is 1153.9 m/g^2, which is enough to settle down 21–38 H atoms, but not enough for a bigger number of H atoms [237].

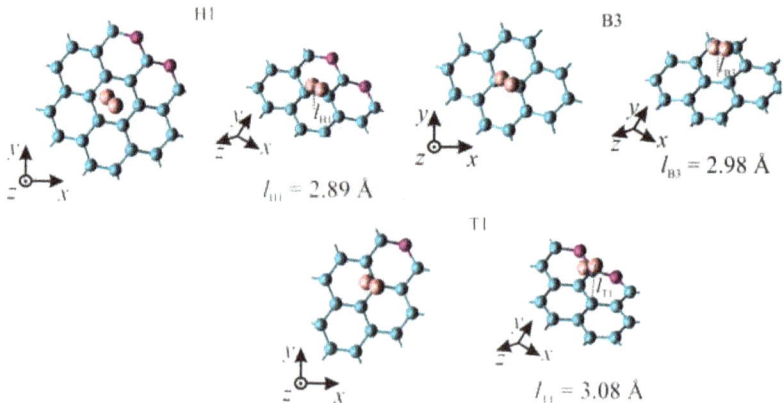

Fig. 5.2. Typical sites of molecular hydrogen on graphene: H1 (hollow), B3 (bridge) and T1 (top). Configurations are presented in two projections: in the plane of the graphene sheet (left) and sloped (right). Reprinted with permission from [137].

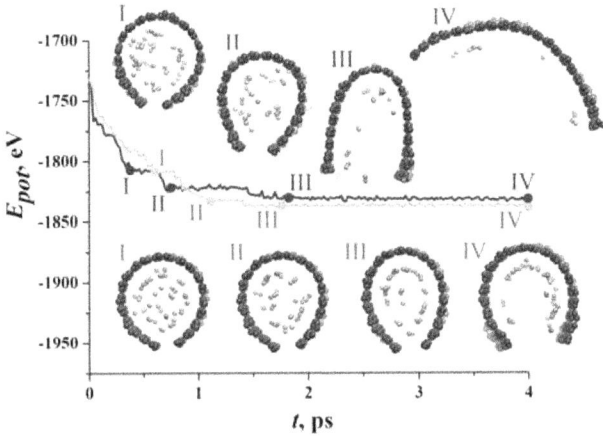

Fig. 5.3. Potential energy as the function of exposure time for hydrogen cluster composed of 47 H atoms. Exposure was conducted at 77 K (gray curve) and 300 K (black curve). Reprinted with permission from [237].

If the hydrogen cluster is quite small in comparison with the specific surface area of the graphene flake, there are a lot of vacant places on the graphene plane to attach hydrogen. Small amounts of hydrogen molecules can move outside the graphene flake at a low temperature equal to 77 K. Most of the hydrogen molecules are attached by van der Waals forces to graphene (physical sorption). At 300 K all hydrogen molecules move outside the graphene flake.

Figure 5.4 presents the results on the hydrogenation of graphene flake in hydrogen atmosphere at two values of applied external pressure equal to 1 and 140 atm. As can be seen from the graphene flake in a hydrogen atmosphere, physical sorption is realized by the van der Waals bonding of H atoms and H_2 molecules, with chemical bonding realized by the formation of CH pairs on the edges of the graphene flake.

The gravimetric density g of hydrogen increases almost monotonously with increasing exposure time. At 1 atm., the value of g after $t = 70$ ps reaches almost a constant value until the end of exposure. The gravimetric density at 140 atm. reaches saturation at $t = 150$ ps. From the snapshots it can be seen that atomic hydrogen adsorbes on the edges of the graphene flake. The higher the external hydrostatic pressure, the more hydrogen atoms and molecules are absorbed by graphene. The maximum value of g is observed at 140 atm. and is about 28 wt.%. At 1 atm. the gravimetric density of hydrogen reaches a constant value of 15 wt.%.

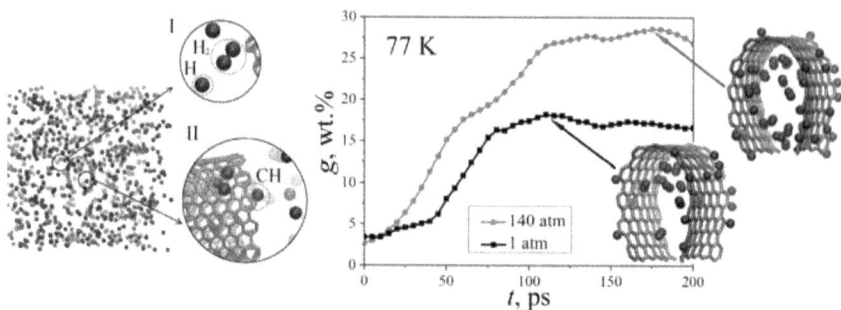

Fig. 5.4. (Left) Graphene flake in a hydrogen atmosphere. Inset (I) shows the atomic hydrogen H and the H_2 molecule, and inset (II) shows the CH group at the edge of the flake. (Right) Gravimetric density as a function of exposure time and snapshots of the graphene flake at 1 atm. (black curve) and 140 atm. (gray curve). Exposure temperature is 77 K. Reprinted with permission from [137].

At 300 K and 140 atm., the number of hydrogen atoms deposited on the flake edge is greater than at 1 atm. The gravimetric density at 1 atm. is about 7 wt.%, and at 140 atm. about 10 wt.% [7]. At 300 K the number of hydrogen molecules and atoms inside the graphene flake is less than at 77 K and the gravimetric density at 300 K is 2.5 times less.

When the hydrogenation temperature decreases from room temperature to 77 K, the physical adsorption of hydrogen at low temperatures becomes more difficult, due to a decrease in the kinetic energy of hydrogen. This energy is not enough for the formation of strong covalent bonds between hydrogen atoms and edge carbon atoms. In contrast, as the temperature decreases, the chemical adsorption of hydrogen molecules increases, increasing gravimetric density. The external applied pressure affects the growth of the gravimetric density of the hydrogen, and the higher the pressure, the more H atoms are adsorbed by the structure. However, an increase in pressure has a greater effect on the increase in the physical adsorption of graphene than chemical.

Curvature of graphene can also considerably affect the hydrogenation. Crumpling is a feasible method to change the graphene sorbability. An increase of adsorption energy at the regions with negative curvature leads to higher diffusion barrier. The difference in adsorption energies for regions with different curvature may be utilized to promote desorption. Figure 5.5 presents two ripple configurations with one and two atomic rows in the ripple bottom and adsorption energies for H on the rippled graphene under compression. The absolute value of the adsorption energy decreased for the

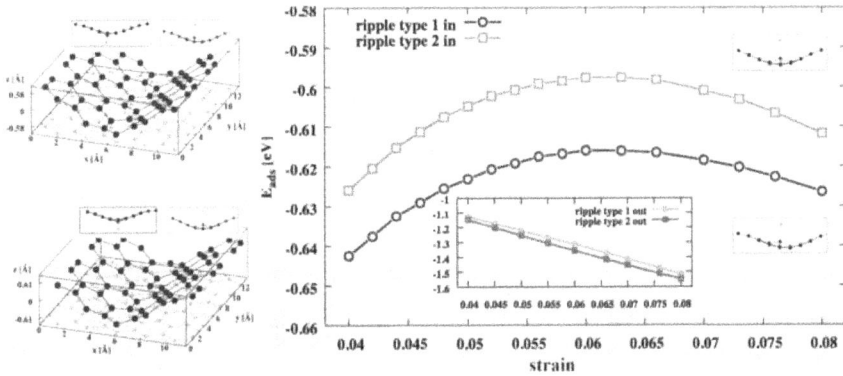

Fig. 5.5. (Left) Two ripple configurations: graphene with ripple of one (top) and two (bottom) atomic rows in the ripple bottom. (Right) Adsorption energies for H inside the rippled graphene under compression. Inset shows energies for H outside the ripple. Reprinted with permission from [159].

"in" atom, and increased for the "out" atom. That means that hydrogen binds with higher energy to the flat graphene than to the regions with positive curvature on the rippled structure. However, regions with negative curvature are even more preferable for H adatom. Change in the ripple configuration lowers the absolute value of the adsorption energy for the second type ripple, making the adsorption site of the first type ripple more attractive for the hydrogen atom [159].

5.2.3 *3D Graphenes*

In order to increase the amount of adsorbed hydrogen, it is necessary to obtain a structure with a large number of micropores, preferentially of the same size. The literature gives various sorption capabilities of the external and internal surfaces of nanotubes, which can be part of a bulk nanomaterial: the internal walls have a higher surface potential for physical sorption of hydrogen. To note, the surface curvature can also affect the surface potential and the amount of adsorbed hydrogen.

Figure 5.6 presents the example of 3D carbon structures which can be used for hydrogen storage and transportation: graphene nanocage, CNT bundle, nanoscroll, pillared graphene, and porous nanotube network. For example, a structure consisting of graphene sheets connected by short CNTs (columnar graphene) are considered very promising especially with Li atoms as the centers for hydrogen storage. Graphene rolls, systems of linked CNTs, crumpled graphene, etc., are also very promising for this field.

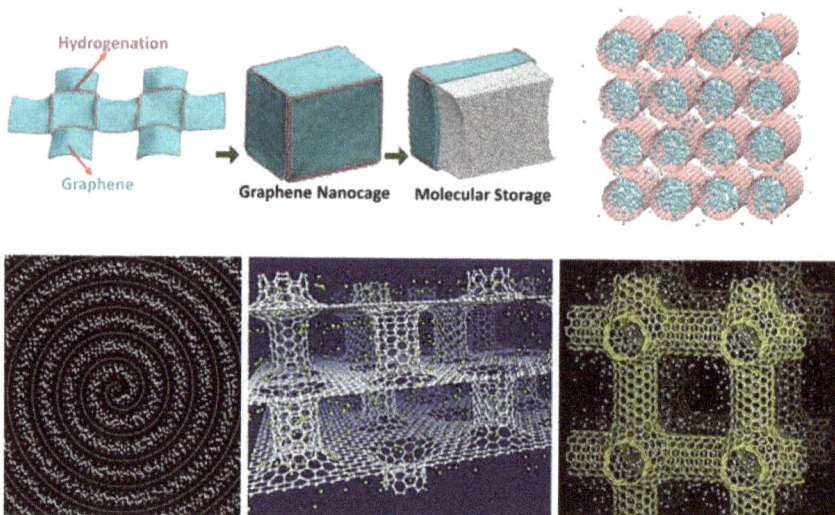

Fig. 5.6. Examples of 3D carbon structures which can be used for hydrogen storage and transportation: graphene nanocage (top left, reprinted with permission from [311]), CNT bundle (top right), nanoscroll (bottom left), pillared graphene (bottom middle), and porous nanotube network (bottom right, reprinted with permission from [75]).

A nanoscroll is a graphene nanoribbon which has a spiral arrangement and can be theoretically obtained by twisting a graphite. It is very similar to multi-walled CNTs, with interlayer distance of approximately 3.6 Å. In this structural state nanoscrolls cannot accumulate enough hydrogen, as the interlayer distance is too small, but an opening of the structure to 7 Å by alkali doping can make them very promising materials for hydrogen storage.

The gravimetric capacity of a material can be affected by the size and chirality of CNTs, as well as temperature, external pressure, and geometry and porosity of the structure. The tunable porosity is the most important aspect and crucial for efficient hydrogen storage. Very small pores cause problems in the insertion of hydrogen molecules, or will not store hydrogen at all.

Figure 5.7 presents one more structure very effective for hydrogen storage – crumpled graphene, which is composed of graphene flakes connected by van der Waals interaction. However, hydrogen can easily move out of pores of this structure. This can be overcomed by hydrostatic compression applied to the hydrogenated crumpled graphene.

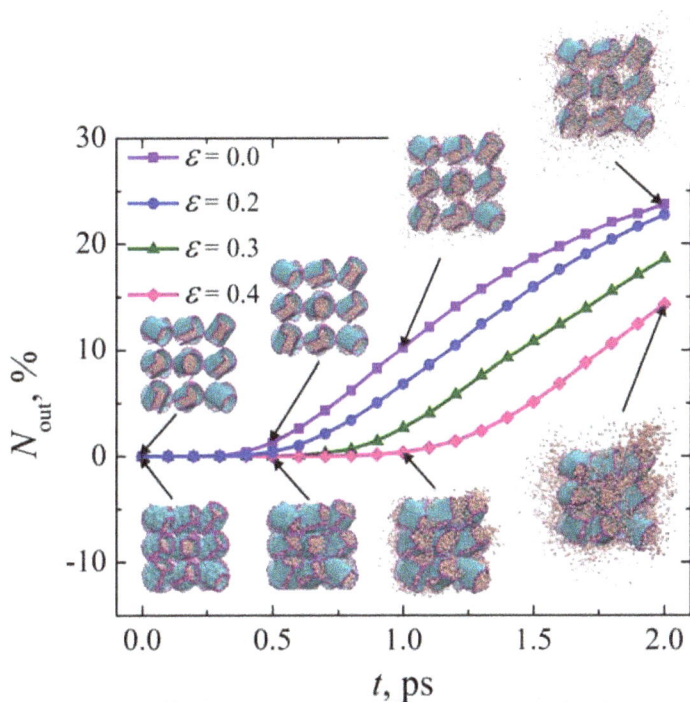

Fig. 5.7. Number of hydrogen atoms (in %) that moved outside crumpled graphene with corresponding snapshots of the structure at 300 K for different values of strain. Reprinted with permission from [137].

Figure 5.7 presents the number of hydrogen atoms (in %) that moved outside crumpled graphene during exposure at 300 K. Snapshots of the structure at different times help to characterize the dehydrogenation process. As can be seen, an increase of deformation until 0.4 leads to a considerable decrease of the number of hydrogen atoms that moved outside the crumpled graphene. At 2 ps, two times fewer hydrogen atoms are displaced from the structure at 0.4 than for the initial structure: 85% of hydrogen is preserved in the compressed crumpled graphene.

Strain increase leads to the considerable growth of volumetric capacity. However, after 2 ps the volumetric capacity of undeformed crumpled graphene becomes higher than that of deformed. It can be explained by the fact that the dehydrogenation of undeformed crumpled graphene goes uniformly, and the hydrogen molecules gradually leave the structure during exposure. In contrast, hydrostatic compression leads to a severe deformation of graphene flakes, due to which hydrogen atoms with a temperature exposure of more than 2 ps leave the structure as a large cluster, and dehydrogenation at this site happens faster. Again, temperature effect is very important. The dehydrogenation of crumpled graphene starts at 50–100 K for undeformed crumpled graphene and at 100–150 K for highest compression 0.4. Further temperature increase results in a sharp decrease of volumetric capacity, but it is still greater than for compressed crumpled graphene up to 500 K. Subsequent rise in temperature does not lead to a significant difference for different values of compression strain. Hydrostatic compression leads to an increase in the starting temperature of the dehydrogenation process of crumpled graphene to higher values.

Initially, there were high expectations of carbon nanostructures becoming an ideal hydrogen carrier. That was supported by preliminary theoretical predictions and experimental studies, which gave the mass fracture of stored hydrogen close to the demand of the International Energy Agency, 5.5 %wt. However, it cannot be easily achieved experimentally and the low hydrogen storage capacity of carbon nanostructures remains a significant shortcoming.

References

[1] Abraham, M. J., Murtola, T., Schulz, R., Páll, S., Smith, J. C., Hess, B., and Lindahl, E. (2015). Gromacs: High performance molecular simulations through multi-level parallelism from laptops to supercomputers, *SoftwareX* **1–2**, pp. 19–25, doi:10.1016/j.softx.2015.06.001.

[2] Akhunova, A., Murzaev, R., and Baimova, J. (2024). Graphene with dislocation dipoles: Wrinkling and defect nucleation during tension, *Computational Materials Science* **244**, p. 113230, doi:10.1016/j.commatsci.2024.113230.

[3] Akinwande, D., Huyghebaert, C., Wang, C.-H., Serna, M. I., Goossens, S., Li, L.-J., Wong, H.-S. P., and Koppens, F. H. L. (2019). Graphene and two-dimensional materials for silicon technology, *Nature* **573**, 7775, pp. 507–518, doi:10.1038/s41586-019-1573-9.

[4] Andrew, R. C., Mapasha, R. E., Ukpong, A. M., and Chetty, N. (2012). Mechanical properties of graphene and boronitrene, *Physical Review B* **85**, 12, p. 125428, doi:10.1103/physrevb.85.125428.

[5] Ansari, R., Mirnezhad, M., and Rouhi, H. (2015). Mechanical properties of fully hydrogenated graphene sheets, *Solid State Communications* **201**, pp. 1–4, doi:10.1016/j.ssc.2014.10.002.

[6] Ansari, R., Sadeghi, F., and Ajori, S. (2013). Continuum and molecular dynamics study of C_{60} fullerene–carbon nanotube oscillators, *Mechanics Research Communications* **47**, pp. 18–23, doi:10.1016/j.mechrescom.2012.11.002.

[7] Apkadirova, N., Krylova, K., and Baimova, J. (2022). Effect of external pressure on the hydrogen storage capacity of a graphene flake: molecular dynamics, *Letters on Materials* **12**, 4s, pp. 445–450, doi:10.22226/2410-3535-2022-4-445-450.

[8] Arsentyev, V. A., Blekhman, I. I., Blekhman, L. I., Vaisberg, L. A., Ivanov, K. S., and Krivtsov, A. M. (2010). Methods of dynamics of

particles and discrete elements as a tool for research and optimization of the processes of processing natural and technogenic materials, *Enrichment of Ruds*, **1**, pp. 30–35.

[9] Asadpour, M., Malakpour, S., Faghihnasiri, M., and Taghipour, B. (2015). Mechanical properties of two-dimensional graphyne sheet, analogous system of BN sheet and graphyne-like BN sheet, *Solid State Communications* **212**, pp. 46–52, doi:10.1016/j.ssc.2015.02. 005.

[10] Bae, S., Kim, H., Lee, Y., Xu, X., Park, J.-S., Zheng, Y., Balakrishnan, J., Lei, T., Ri Kim, H., Song, Y. I., Kim, Y.-J., Kim, K. S., Ozyilmaz, B., Ahn, J.-H., Hong, B. H., and Iijima, S. (2010). Roll-to-roll production of 30-inch graphene films for transparent electrodes, *Nature Nanotechnology* **5**, 8, pp. 574–578, doi:10.1038/nnano.2010. 132.

[11] Baimova, J. A., Liu, B., Dmitriev, S. V., Srikanth, N., and Zhou, K. (2014). Mechanical properties of bulk carbon nanostructures: effect of loading and temperature, *Physical Chemistry Chemical Physics* **16**, pp. 19505–19513, doi:10.1039/C4CP01952K.

[12] Baimova, J. A. (2024). An overview of mechanical properties of diamond-like phases under tension, *Nanomaterials* **14**, 2, p. 129, doi: 10.3390/nano14020129.

[13] Baimova, J. A. and Shcherbinin, S. A. (2022). Metal/graphene composites: A review on the simulation of fabrication and study of mechanical properties, *Materials* **16**, 1, p. 202, doi:10.3390/ ma16010202.

[14] Baimova, J. A., Dmitriev, S. V., and Zhou, K. (2012). Strain-induced ripples in graphene nanoribbons with clamped edges, *physica status solidi (b)* **249**, 7, pp. 1393–1398, doi:10.1002/pssb. 201084224.

[15] Baimova, J. A., Dmitriev, S. V., Zhou, K., and Savin, A. V. (2012). Unidirectional ripples in strained graphene nanoribbons with clamped edges at zero and finite temperatures, *Physical Review B* **86**, 3, p. 035427, doi:10.1103/physrevb.86.035427.

[16] Bakharev, P. V., Huang, M., Saxena, M., Lee, S. W., Joo, S. H., Park, S. O., Dong, J., Camacho-Mojica, D. C., Jin, S., Kwon, Y., Biswal, M., Ding, F., Kwak, S. K., Lee, Z., and Ruoff, R. S. (2019). Chemically induced transformation of chemical vapour deposition grown bilayer graphene into fluorinated single-layer diamond, *Nature Nanotechnology* **15**, 1, pp. 59–66, doi:10.1038/s41565-019-0582-z.

[17] Balandin, A. A., Ghosh, S., Bao, W., Calizo, I., Teweldebrhan, D., Miao, F., and Lau, C. N. (2008). Superior thermal conductivity of single-layer graphene, *Nano Letters* **8**, 3, pp. 902–907, doi:10.1021/ nl0731872.

[18] Banhart, F., Redlich, P., and Ajayan, P. (1998). The migration of metal atoms through carbon onions, *Chemical Physics Letters* **292**, 4–6, pp. 554–560, doi:10.1016/s0009-2614(98)00705-2.

[19] Bao, W., Miao, F., Chen, Z., Zhang, H., Jang, W., Dames, C., and Lau, C. N. (2009). Controlled ripple texturing of suspended graphene and ultrathin graphite membranes, *Nature Nanotechnology* **4**, 9, pp. 562–566, doi:10.1038/nnano.2009.191.

[20] Baughman, R. H., Eckhardt, H., and Kertesz, M. (1987). Structure-property predictions for new planar forms of carbon: Layered phases containing sp^2 and sp atoms, *The Journal of Chemical Physics* **87**, 11, pp. 6687–6699, doi:10.1063/1.453405.

[21] Bejagam, K. K., Singh, S., and Deshmukh, S. A. (2018). Nanoparticle activated and directed assembly of graphene into a nanoscroll, *Carbon* **134**, pp. 43–52, doi:10.1016/j.carbon.2018.03.077.

[22] Belenkov, E. A., Mavrinskii, V. V., Belenkova, T. E., and Chernov, V. M. (2015). Structural modifications of graphyne layers consisting of carbon atoms in the sp- and sp^2-hybridized states, *Journal of Experimental and Theoretical Physics* **120**, 5, pp. 820–830, doi:10.1134/s1063776115040214.

[23] Berinskii, I. E. and Krivtsov, A. M. (2010). On using many-particle interatomic potentials to compute elastic properties of graphene and diamond, *Mechanics of Solids* **45**, 6, pp. 815–834, doi:10.3103/s0025654410060063.

[24] Bhattacharya, B., Singh, N. B., and Sarkar, U. (2015). Pristine and BN doped graphyne derivatives for UV light protection, *International Journal of Quantum Chemistry* **115**, 13, pp. 820–829, doi:10.1002/qua.24910.

[25] Biris, A. R., Lazar, M. D., Pruneanu, S., Neamtu, C., Watanabe, F., Kannarpady, G. K., Dervishi, E., and Biris, A. S. (2013). Catalytic one-step synthesis of Pt-decorated few-layer graphenes, *RSC Advances* **3**, 48, p. 26391, doi:10.1039/c3ra44564j.

[26] Bochvar, D. A. and Galpern, E. G. (1973). On hypothetical systems: carbododecahedron, s-icosahedron and carbo-s-icosahedron, *Reports of the USSR Academy of Sciences* **209**, 3, pp. 610–612.

[27] Bogdanova, A. R., Krasnikov, D. V., and Nasibulin, A. G. (2023). The role of sulfur in the CVD carbon nanotube synthesis, *Carbon* **210**, p. 118051, doi:10.1016/j.carbon.2023.118051.

[28] Bogdanova, A. R., Krasnikov, D. V., Khabushev, E. M., Ramirez B., J. A., Matyushkin, Y. E., and Nasibulin, A. G. (2023). Role of hydrogen in ethylene-based synthesis of single-walled carbon nanotubes, *Nanomaterials* **13**, 9, p. 1504, doi:10.3390/nano13091504.

[29] Bolotin, K., Sikes, K., Jiang, Z., Klima, M., Fudenberg, G., Hone, J., Kim, P., and Stormer, H. (2008). Ultrahigh electron mobility

in suspended graphene, *Solid State Communications* **146**, 9–10, pp. 351–355, doi:10.1016/j.ssc.2008.02.024.

[30] Bonaccorso, F., Colombo, L., Yu, G., Stoller, M., Tozzini, V., Ferrari, A. C., Ruoff, R. S., and Pellegrini, V. (2015). Graphene, related two-dimensional crystals, and hybrid systems for energy conversion and storage, *Science* **347**, 6217, doi:10.1126/science.1246501.

[31] Bonilla, L. L. and Carpio, A. (2012). Driving dislocations in graphene, *Science* **337**, 6091, pp. 161–162, doi:10.1126/science. 1224681.

[32] Brenner, D. W. (1990). Empirical potential for hydrocarbons for use in simulating the chemical vapor deposition of diamond films, *Physical Review B* **42**, 15, pp. 9458–9471, doi:10.1103/physrevb.42. 9458.

[33] Brooks, B. R., Bruccoleri, R. E., Olafson, B. D., States, D. J., Swaminathan, S., and Karplus, M. (1983). Charmm: A program for macromolecular energy, minimization, and dynamics calculations, *Journal of Computational Chemistry* **4**, 2, pp. 187–217, doi:10.1002/ jcc.540040211.

[34] Bühl, M. (1998). The relation between endohedral chemical shifts and local aromaticities in fullerenes, *Chemistry - A European Journal* **4**, 4, pp. 734–739, doi:10.1002/(sici)1521-3765(19980416)4:4⟨734:: aid-chem734⟩3.0.co;2-c.

[35] Cabanillas-Casas, C. J., Cuba-Supanta, G., Rojas Tapia, J., and Rojas-Ayala, C. (2021). Efectos de forma y tamaño del poro sobre las propiedades mecánicas de las membranas del grafeno nanoporoso, *MOMENTO*, 62, pp. 63–78, doi:10.15446/mo.n62.88422.

[36] Cai, W., Moore, A. L., Zhu, Y., Li, X., Chen, S., Shi, L., and Ruoff, R. S. (2010). Thermal transport in suspended and supported monolayer graphene grown by chemical vapor deposition, *Nano Letters* **10**, 5, pp. 1645–1651, doi:10.1021/nl9041966.

[37] Cao, C., Sun, Y., and Filleter, T. (2014). Characterizing mechanical behavior of atomically thin films: A review, *Journal of Materials Research* **29**, 3, pp. 338–347, doi:10.1557/jmr.2013.339.

[38] Case, D. A., Cheatham, T. E., Darden, T., Gohlke, H., Luo, R., Merz, K. M., Onufriev, A., Simmerling, C., Wang, B., and Woods, R. J. (2005). The amber biomolecular simulation programs, *Journal of Computational Chemistry* **26**, 16, pp. 1668–1688, doi:10.1002/jcc. 20290.

[39] Castro Neto, A. H., Guinea, F., Peres, N. M. R., Novoselov, K. S., and Geim, A. K. (2009). The electronic properties of graphene, *Reviews of Modern Physics* **81**, pp. 109–162, doi:10.1103/RevModPhys.81.109.

[40] Cellini, F., Lavini, F., Cao, T., de Heer, W., Berger, C., Bongiorno, A., and Riedo, E. (2018). Epitaxial two-layer graphene under pressure: Diamene stiffer than diamond, *FlatChem* **10**, pp. 8–13, doi:10.1016/j.flatc.2018.08.001.

[41] Chadli, H., Rahmani, A., and Sauvajol, J.-L. (2010). Raman spectra of C_{60} dimer and C_{60} polymer confined inside a (10, 10) single-walled carbon nanotube, *Journal of Physics: Condensed Matter* **22**, 14, p. 145303, doi:10.1088/0953-8984/22/14/145303.

[42] Chadli, H., Rahmani, A., Sbai, K., Hermet, P., Rols, S., and Sauvajol, J.-L. (2006). Calculation of Raman-active modes in linear and zigzag phases of fullerene peapods, *Physical Review B* **74**, 20, p. 205412, doi:10.1103/physrevb.74.205412.

[43] Chen, Y. L., Liu, B., Wu, J., Huang, Y., Jiang, H., and Hwang, K. C. (2008). Mechanics of hydrogen storage in carbon nanotubes, *Journal of the Mechanics and Physics of Solids* **56**, 11, pp. 3224–3241.

[44] Chen, Z., Wang, X., Bhakhri, V., Giuliani, F., and Atkinson, A. (2013). Nanoindentation of porous bulk and thin films of $La_{0.6}Sr_{0.4}Co_{0.2}Fe_{0.8}O_3$, *Acta Materialia* **61**, 15, pp. 5720–5734, doi:10.1016/j.actamat.2013.06.016.

[45] Cheng, Y., Zhou, S., Hu, P., Zhao, G., Li, Y., Zhang, X., and Han, W. (2017). Enhanced mechanical, thermal, and electric properties of graphene aerogels via supercritical ethanol drying and high-temperature thermal reduction, *Scientific Reports* **7**, pp. 1439, doi:10.1038/s41598-017-01601-x.

[46] Chernozatonskii, L. A., Katin, K. P., Demin, V. A., and Maslov, M. M. (2021). Moiré diamanes based on the hydrogenated or fluorinated twisted bigraphene: The features of atomic and electronic structures, Raman and infrared spectra, *Applied Surface Science* **537**, p. 148011, doi:10.1016/j.apsusc.2020.148011.

[47] Chernozatonskii, L. A., Sorokin, P. B., Kuzubov, A. A., Sorokin, B. P., Kvashnin, A. G., Kvashnin, D. G., Avramov, P. V., and Yakobson, B. I. (2010). Influence of size effect on the electronic and elastic properties of diamond films with nanometer thickness, *The Journal of Physical Chemistry C* **115**, 1, pp. 132–136, doi:10.1021/jp1080687.

[48] Chernozatonskii, L. A., Sorokin, P. B., Kvashnin, A. G., and Kvashnin, D. G. (2009). Diamond-like C_{2h} nanolayer, diamane: Simulation of the structure and properties, *JETP Letters* **90**, 2, pp. 134–138, doi:10.1134/s0021364009140112.

[49] Choi, B. K., Yoon, G. H., and Lee, S. (2016). Molecular dynamics studies of CNT-reinforced aluminum composites under uniaxial

tensile loading, *Composites Part B: Engineering* **91**, pp. 119–125, doi:10.1016/j.compositesb.2015.12.031.

[50] Chu, Y., Ragab, T., Gautreau, P., and Basaran, C. (2015). Mechanical properties of hydrogen edge–passivated chiral graphene nanoribbons, *Journal of Nanomechanics and Micromechanics* **5**, 4, doi: 10.1061/(asce)nm.2153-5477.0000101.

[51] Coluci, V. R., Braga, S. F., Legoas, S. B., Galvão, D. S., and Baughman, R. H. (2003). Families of carbon nanotubes: Graphyne-based nanotubes, *Physical Review B* **68**, 3, p. 035430, doi:10.1103/physrevb.68.035430.

[52] Cranford, S. W. and Buehler, M. J. (2011). Mechanical properties of graphyne, *Carbon* **49**, 13, pp. 4111–4121, doi:10.1016/j.carbon.2011.05.024.

[53] Cranford, S. W., Brommer, D. B., and Buehler, M. J. (2012). Extended graphynes: simple scaling laws for stiffness, strength and fracture, *Nanoscale* **4**, 24, p. 7797, doi:10.1039/c2nr31644g.

[54] Deng, S. and Berry, V. (2016). Wrinkled, rippled and crumpled graphene: an overview of formation mechanism, electronic properties, and applications, *Materials Today* **19**, 4, pp. 197–212, doi:10.1016/j.mattod.2015.10.002.

[55] Desyatkin, V. G., Martin, W. B., Aliev, A. E., Chapman, N. E., Fonseca, A. F., Galvao, D. S., Miller, E. R., Stone, K. H., Wang, Z., Zakhidov, D., Limpoco, F. T., Almahdali, S. R., Parker, S. M., Baughman, R. H., and Rodionov, V. O. (2022). Scalable synthesis and characterization of multilayer gamma-graphyne, new carbon crystals with a small direct band gap, *Journal of the American Chemical Society* **144**, 39, pp. 17999–18008, doi:10.1021/jacs.2c06583.

[56] Dewapriya, M. A. N. and Rajapakse, R. K. N. D. (2014). Molecular dynamics simulations and continuum modeling of temperature and strain rate dependent fracture strength of graphene with vacancy defects, *Journal of Applied Mechanics* **81**, 8, p. 081010, doi:10.1115/1.4027681.

[57] Diederich, F. (1994). Carbon scaffolding: building acetylenic all-carbon and carbon-rich compounds, *Nature* **369**, 6477, pp. 199–207, doi:10.1038/369199a0.

[58] Diederich, F. and Rubin, Y. (1992). Synthetic approaches toward molecular and polymeric carbon allotropes, *Angewandte Chemie International Edition in English* **31**, 9, pp. 1101–1123, doi:10.1002/anie.199211013.

[59] Ding, Y., Li, D., Xu, F., Lang, W., Qin, Q. H., Ye, Z., Liu, J., and Wen, X. (2024). The microstructure evolution of graphene in nanoindentation G/WC-Co based on molecular dynamics simulation,

Diamond and Related Materials **141**, p. 110729, doi:10.1016/j. diamond.2023.110729.

[60] Di Giorgio, C., Blundo, E., Pettinari, G., Felici, M., Bobba, F., and Polimeni, A. (2022). Mechanical, elastic, and adhesive properties of two-dimensional materials: From straining techniques to state-of-the-art local probe measurements, *Advanced Materials Interfaces* **9**, 13, p. 2102220, doi:10.1002/admi.202102220.

[61] Du, X., Skachko, I., Barker, A., and Andrei, E. Y. (2008). Approaching ballistic transport in suspended graphene, *Nature Nanotechnology* **3**, 8, pp. 491–495, doi:10.1038/nnano.2008.199.

[62] Du, Y.-X., Zhou, L.-J., and Guo, J.-G. (2022). The influence of strain range, size and chiral on mechanical properties of graphene: Molecular dynamics insights, *Nanomaterials and Nanotechnology* **12**, p. 184798042211100, doi:10.1177/18479804221110023.

[63] Duan, W. H., Gong, K., and Wang, Q. (2011). Controlling the formation of wrinkles in a single layer graphene sheet subjected to in-plane shear, *Carbon* **49**, 9, pp. 3107–3112, doi:10.1016/j.carbon. 2011.03.033.

[64] Duhan, N. and Kumar, T. D. (2024). Ab initio study of Li-shrouded Si-doped γ-graphyne nanosheet as propitious anode in Li-ion batteries, *Applied Surface Science* **642**, p. 158553, doi:10.1016/ j.apsusc.2023.158553.

[65] El Goresy, A. and Donnay, G. T. (1968). A new allotropic form of carbon from the Ries Crater. *Science* **161**, pp. 363–364, doi:10.1126/ science.161.3839.363.

[66] Elias, D. C., Nair, R. R., Mohiuddin, T. M. G., Morozov, S. V., Blake, P., Halsall, M. P., Ferrari, A. C., Boukhvalov, D. W., Katsnelson, M. I., Geim, A. K., and Novoselov, K. S. (2009). Control of graphene's properties by reversible hydrogenation: Evidence for graphane, *Science* **323**, 5914, pp. 610–613, doi:10.1126/science.1167130.

[67] Entani, S., Antipina, L. Y., Avramov, P. V., Ohtomo, M., Matsumoto, Y., Hirao, N., Shimoyama, I., Naramoto, H., Baba, Y., Sorokin, P. B., and Sakai, S. (2015). Contracted interlayer distance in graphene/sapphire heterostructure, *Nano Research* **8**, 5, pp. 1535–1545, doi:10.1007/s12274-014-0640-7.

[68] Erohin, S. V., Ruan, Q., Sorokin, P. B., and Yakobson, B. I. (2020). Nano-thermodynamics of chemically induced graphene–diamond transformation, *Small* **16**, 47, p. 2004782, doi:10.1002/smll. 202004782.

[69] Falin, A., Cai, Q., Santos, E. J., Scullion, D., Qian, D., Zhang, R., Yang, Z., Huang, S., Watanabe, K., Taniguchi, T., Barnett, M. R., Chen, Y., Ruoff, R. S., and Li, L. H. (2017). Mechanical properties of

atomically thin boron nitride and the role of interlayer interactions, *Nature Communications* **8**, 1, p. 15815, doi:10.1038/ncomms15815.

[70] Fasolino, A., Los, J. H., and Katsnelson, M. I. (2007). Intrinsic ripples in graphene, *Nature Materials* **6**, 11, pp. 858–861, doi: 10.1038/nmat2011.

[71] Fei, Y., Fang, S., and Hu, Y. H. (2020). Synthesis, properties and potential applications of hydrogenated graphene, *Chemical Engineering Journal* **397**, p. 125408, doi:10.1016/j.cej.2020.125408.

[72] Feynman, R. P. (1960). There's plenty of room at the bottom, *Eng. Sci.* **23**, p. 22–36.

[73] Finnis, M. W. and Sinclair, J. E. (1984). A simple empirical N-body potential for transition metals, *Philosophical Magazine A* **50**, pp. 45–55, doi:10.1080/01418618408244210.

[74] Frank, I. W., Tanenbaum, D. M., van der Zande, A. M., and McEuen, P. L. (2007). Mechanical properties of suspended graphene sheets, *Journal of Vacuum Science and Technology B: Microelectronics and Nanometer Structures Processing, Measurement, and Phenomena* **25**, 6, pp. 2558–2561, doi:10.1116/1.2789446.

[75] Froudakis, G. E. (2011). Hydrogen storage in nanotubes & nanostructures, *Materials Today* **14**, 7–8, pp. 324–328, doi:10.1016/s1369-7021(11)70162-6.

[76] Galashev, A. Y. and Rakhmanova, O. R. (2020). Computational study of the formation of aluminum-graphene nanocrystallites, *Physics Letters A* **384**, 31, p. 126790, doi:10.1016/j.physleta.2020.126790.

[77] Galashev, A. Y., Katin, K. P., and Maslov, M. M. (2019). Morse parameters for the interaction of metals with graphene and silicene, *Physics Letters A* **383**, 2-3, pp. 252–258, doi:https://doi.org/10.1016/j.physleta.2018.10.025.

[78] Galiakhmetova, L., Safina, L., Murzaev, R., and Baimova, J. (2024). Dynamics of dislocation dipoles in graphene at high temperatures, *Diamond and Related Materials* **143**, p. 110896, doi:10.1016/j.diamond.2024.110896.

[79] Gan, Y. and Banhart, F. (2008). The mobility of carbon atoms in graphitic nanoparticles studied by the relaxation of strain in carbon onions, *Advanced Materials* **20**, 24, pp. 4751–4754, doi:10.1002/adma.200800574.

[80] Gao, Y., Cao, T., Cellini, F., Berger, C., de Heer, W. A., Tosatti, E., Riedo, E., and Bongiorno, A. (2017). Ultrahard carbon film from epitaxial two-layer graphene, *Nature Nanotechnology* **13**, 2, pp. 133–138, doi:10.1038/s41565-017-0023-9.

[81] Ge, L., Liu, H., Wang, J., Huang, H., Cui, Z., Huang, Q., Fu, Z., and Lu, Y. (2021). Properties of diamane anchored with different groups, *Physical Chemistry Chemical Physics* **23**, 26, pp. 14195–14204, doi: 10.1039/d1cp01747k.

[82] Geim, A. K. (2009). Graphene: Status and prospects, *Science* **324**, 5934, pp. 1530–1534, doi:10.1126/science.1158877.

[83] Geim, A. K. and Novoselov, K. S. (2007). The rise of graphene, *Nature Materials* **6**, 3, pp. 183–191, doi:10.1038/nmat1849.

[84] Ghosh, S., Bao, W., Nika, D. L., Subrina, S., Pokatilov, E. P., Lau, C. N., and Balandin, A. A. (2010). Dimensional crossover of thermal transport in few-layer graphene, *Nature Materials* **9**, 7, pp. 555–558, doi:10.1038/nmat2753.

[85] Ghosh, S., Calizo, I., Teweldebrhan, D., Pokatilov, E. P., Nika, D. L., Balandin, A. A., Bao, W., Miao, F., and Lau, C. N. (2008). Extremely high thermal conductivity of graphene: Prospects for thermal management applications in nanoelectronic circuits, *Applied Physics Letters* **92**, 15, p. 151911, doi:10.1063/1.2907977.

[86] Gil, A. J., Adhikari, S., Scarpa, F., and Bonet, J. (2010). The formation of wrinkles in single-layer graphene sheets under nanoindentation, *Journal of Physics: Condensed Matter* **22**, 14, p. 145302, doi:10.1088/0953-8984/22/14/145302.

[87] Giovannetti, G., Khomyakov, P. A., Brocks, G., Karpan, V. M., van den Brink, J., and Kelly, P. J. (2008). Doping graphene with metal contacts, *Physical Review Letters* **101**, 2, p. 026803, doi:10.1103/physrevlett.101.026803.

[88] Girifalco, L. A. and Weizer, V. G. (1959). Application of the Morse potential function to cubic metals, *Physical Review* **114**, 3, pp. 687–690, doi:10.1103/PhysRev.114.687.

[89] Gong, C. and Zhang, X. (2019). Two-dimensional magnetic crystals and emergent heterostructure devices, *Science* **363**, 6428, p. eaav4450, doi:10.1126/science.aav4450.

[90] Grüneis, A., Saito, R., Kimura, T., Cancado, L. G., Pimenta, M. A., Jorio, A., Souza Filho, A. G., Dresselhaus, G., and Dresselhaus, M. S. (2002). Probing the phonon dispersion relations of graphite from the double-resonance process of Stokes and anti-Stokes Raman scatterings in multiwalled carbon nanotubes, *Physical Review B* **65**, p. 155405, doi:10.1103/PhysRevB.65.155405.

[91] Ho, D. T., Ho, V. H., Babar, V., Kim, S. Y., and Schwingenschlögl, U. (2020). Complex three-dimensional graphene structures driven by surface functionalization, *Nanoscale* **12**, 18, pp. 10172–10179, doi: 10.1039/D0NR01733G.

[92] Grishakov, K., Merinov, V., Katin, K., and Maslov, M. (2025). Electronic characteristics of diamanes and diamane-based hetero-junctions, *Physica E: Low-dimensional Systems and Nanostructures* **171**, p. 116248, doi:10.1016/j.physe.2025.116248.

[93] Grishakov, K. S., Katin, K. P., and Maslov, M. M. (2018). Strain-induced semiconductor-to-metal transitions in C_{36}-based carbon peapods: Ab initio study, *Diamond and Related Materials* **84**, pp. 112–118, doi:10.1016/j.diamond.2018.03.023.

[94] Haley, M. M., Brand, S. C., and Pak, J. J. (1997). Carbon networks based on dehydrobenzoannulenes: Synthesis of graphdiyne substruc-tures, *Angewandte Chemie International Edition in English* **36**, 8, pp. 836–838, doi:10.1002/anie.199708361.

[95] Hamada, I. and Otani, M. (2010). Comparative van der Waals density-functional study of graphene on metal surfaces, *Physical Review B* **82**, 15, p. 153412, doi:10.1103/physrevb.82.153412.

[96] Harris, P. J. F. (2003). *Uglerodnye nanotruby i rodstvennye struktury. Novye materialy XXI veka.* (Cambridge University Press).

[97] He, R., Zhao, L., Petrone, N., Kim, K. S., Roth, M., Hone, J., Kim, P., Pasupathy, A., and Pinczuk, A. (2012). Large physisorption strain in chemical vapor deposition of graphene on copper substrates, *Nano Letters* **12**, 5, pp. 2408–2413, doi:10.1021/nl300397v.

[98] Hernandez, E., Meunier, V., Smith, B. W., Rurali, R., Terrones, H., Buongiorno Nardelli, M., Terrones, M., Luzzi, D. E., and Charlier, J.-C. (2003). Fullerene coalescence in nanopeapods: a path to novel tubular carbon, *Nano Letters* **3**, 8, pp. 1037–1042, doi:10.1021/nl034283f.

[99] Hernandez, S. A. and Fonseca, A. F. (2017). Anisotropic elastic modulus, high Poisson's ratio and negative thermal expansion of graphynes and graphdiynes, *Diamond and Related Materials* **77**, pp. 57–64, doi:10.1016/j.diamond.2017.06.002.

[100] Hodak, M. and Girifalco, L. A. (2003). Ordered phases of fullerene molecules formed inside carbon nanotubes, *Physical Review B* **67**, 7, p. 075419, doi:10.1103/physrevb.67.075419.

[101] Hodak, M. and Girifalco, L. A. (2003). Systems of C_{60} molecules inside (10, 10) and (15, 15) nanotube: A Monte Carlo study, *Physical Review B* **68**, 8, p. 085405, doi:10.1103/physrevb.68.085405.

[102] Hoover, W. G. (1985). Canonical dynamics: Equilibrium phase-space distributions, *Physical Review A* **31**, 3, pp. 1695–1697, doi:10.1103/physreva.31.1695.

[103] Hornbaker, D. J., Kahng, S.-J., Misra, S., Smith, B. W., Johnson, A. T., Mele, E. J., Luzzi, D. E., and Yazdani, A. (2002). Mapping the one-dimensional electronic states of nanotube peapod structures, *Science* **295**, 5556, pp. 828–831, doi:10.1126/science.1068133.

[104] Hou, X., Xie, Z., Li, C., Li, G., and Chen, Z. (2018). Study of electronic structure, thermal conductivity, elastic and optical properties of alpha, beta, gamma-graphyne, *Materials* **11**, 2, p. 188, doi:10.3390/ma11020188.

[105] Houshmand, F., Jalili, S., and Schofield, J. (2016). Halogenated graphdiyne and graphyne single layers: A systematic study, *Physical Chemistry Research* **4**, 2, pp. 231–243, doi:10.22036/pcr.2016.13940.

[106] Hu, M., He, J., Wang, Q., Huang, Q., Yu, D., Tian, Y., and Xu, B. (2014). Covalent-bonded graphyne polymers with high hardness, *Journal of Superhard Materials* **36**, 4, pp. 257–269, doi:10.3103/s1063457614040042.

[107] Hu, Y., Li, D., Yin, Y., Li, S., Ding, G., Zhou, H., and Zhang, G. (2020). The important role of strain on phonon hydrodynamics in diamond-like bi-layer graphene, *Nanotechnology* **31**, 33, p. 335711, doi:10.1088/1361-6528/ab8ee1.

[108] Huang, H., Tang, X., Chen, F., Liu, J., and Chen, D. (2017). Role of graphene layers on the radiation resistance of copper–graphene nanocomposite: Inhibiting the expansion of thermal spike, *Journal of Nuclear Materials* **493**, pp. 322–329, doi:10.1016/j.jnucmat.2017.06.023.

[109] Huang, X., Zhao, G., and Wang, X. (2015). Fabrication of reduced graphene oxide/metal (Cu, Ni, Co) nanoparticle hybrid composites via a facile thermal reduction method, *RSC Advances* **5**, 62, pp. 49973–49978, doi:10.1039/c5ra08670a.

[110] Iijima, S. (1991). Helical microtubules of graphitic carbon, *Nature* **354**, 6348, pp. 56–58, doi:10.1038/354056a0.

[111] Iyoda, M., Ishita, M., Ohkoshi, M., Kuwatani, Y., Otani, H., and Nishinaga, T. (2019). Synthesis and properties of a tricyclic hexaketone monohydrate with hexabutyl side chain, *HETEROCYCLES* **99**, 2, p. 1145, doi:10.3987/com-18-s(f)85.

[112] Jacob, W. and Möller, W. (1993). On the structure of thin hydrocarbon films, *Applied Physics Letters* **63**, 13, pp. 1771–1773, doi:10.1063/1.110683.

[113] Jin, H., Guo, C., Liu, X., Liu, J., Vasileff, A., Jiao, Y., Zheng, Y., and Qiao, S.-Z. (2018). Emerging two-dimensional nanomaterials for electrocatalysis, *Chemical Reviews* **118**, 13, pp. 6337–6408, doi:10.1021/acs.chemrev.7b00689.

[114] Jo, E. H., Choi, J.-H., Park, S.-R., Lee, C. M., Chang, H., and Jang, H. D. (2016). Size and structural effect of crumpled graphene balls on the electrochemical properties for supercapacitor application, *Electrochimica Acta* **222**, pp. 58–63, doi:10.1016/j.electacta.2016.11.016.

[115] Kanegae, G. B. and Fonseca, A. F. (2022). Effective acetylene length dependence of the elastic properties of different kinds of graphynes, *Carbon Trends* **7**, p. 100152, doi:10.1016/j.cartre.2022.100152.

[116] Kang, J., Li, J., Wu, F., Li, S.-S., and Xia, J.-B. (2011). Elastic, electronic, and optical properties of two-dimensional graphyne sheet, *The Journal of Physical Chemistry C* **115**, 42, pp. 20466–20470, doi: 10.1021/jp206751m.

[117] Kasahara, Y., Takeuchi, Y., Zadik, R. H., Takabayashi, Y., Colman, R. H., McDonald, R. D., Rosseinsky, M. J., Prassides, K., and Iwasa, Y. (2017). Upper critical field reaches 90 Tesla near the Mott transition in fulleride superconductor, *Nature Communications* **8**, 1, p. 14467, doi:10.1038/ncomms14467.

[118] Kasatochkin, V. I., Sladkov, A. M., Kudryavtsev, Y. P., and Korshak, V. V. (1969). *Structural Chemistry of Carbon and Coals*. Nauka, Moscow; US Department of the Interior Bureau of Mines and National Science Foundation, Washington, DC, United States (1976).

[119] Katin, K., Kaya, S., and Maslov, M. (2022). Graphene nanoflakes and fullerenes doped with aluminum: features of Al-C interaction and adsorption characteristics of carbon shell, *Letters on Materials* **12**, 2, pp. 148–152, doi:10.22226/2410-3535-2022-2-148-152.

[120] Katin, K. P., Podlivaev, A. I., Kochaev, A. I., Kulyamin, P. A., Bauetdinov, Y., Grekova, A. A., Bereznitskiy, I. V., and Maslov, M. M. (2024). Diamanes from novel graphene allotropes: Computational study on structures, stabilities and properties, *FlatChem* **44**, p. 100622, doi:10.1016/j.flatc.2024.100622.

[121] Katin, K. P., Prudkovskiy, V. S., and Maslov, M. M. (2018). Molecular dynamics simulation of nickel-coated graphene bending, *Micro & Nano Letters* **13**, 2, pp. 160–164, doi:10.1049/mnl.2017.0460.

[122] Kawamura, Y. and Ohta, Y. (2022). Annihilation dynamics of a dislocation pair in graphene: Density-functional tight-binding molecular dynamics simulations and first principles study, *Computational Materials Science* **205**, p. 111224, doi:10.1016/j.commatsci.2022.111224.

[123] Khlobystov, A. N., Britz, D. A., Ardavan, A., and Briggs, G. A. D. (2004). Observation of ordered phases of fullerenes in carbon nanotubes, *Physical Review Letters* **92**, 24, p. 245507, doi:10.1103/physrevlett.92.245507.

[124] Kim, K. S., Zhao, Y., Jang, H., Lee, S. Y., Kim, J. M., Kim, K. S., Ahn, J.-H., Kim, P., Choi, J.-Y., and Hong, B. H. (2009). Large-scale pattern growth of graphene films for stretchable transparent electrodes, *Nature* **457**, 7230, pp. 706–710, doi:10.1038/nature07719.

[125] Kittel, C. (2004). *Introduction to Solid State Physics* (Wiley).

[126] Kochaev, A. I., Efimov, V. V., Kaya, S., Flores-Moreno, R., Katin, K. P., and Maslov, M. M. (2023). On point perforating defects in bilayer structures, *Physical Chemistry Chemical Physics* **25**, 44, pp. 30477–30487, doi:10.1039/d3cp03719c.

[127] Kolesnikov, V., Mironov, R., and Baimova, J. (2024). Graphene/ Metal composites decorated with Ni nanoclusters: Mechanical properties, *Materials* **17**, 23, p. 5753, doi:10.3390/ma17235753.

[128] Kotakoski, J., Krasheninnikov, A. V., Kaiser, U., and Meyer, J. C. (2011). From point defects in graphene to two-dimensional amorphous carbon, *Physical Review Letters* **106**, 10, p. 105505, doi: 10.1103/physrevlett.106.105505.

[129] Krasnikov, D. V., Marunchenko, A. A., Koroleva, E. A., Kondrashov, V. A., Ilatovskii, D. A., Khabushev, E. M., Dmitrieva, V. A., Iakovlev, V. Y., Kopylova, D. S., Baklanov, A. M., Shandakov, S. D., and Nasibulin, A. G. (2024). One-step dry deposition technique for aligning single-walled carbon nanotubes, *Chemical Engineering Journal* **498**, p. 155508, doi:10.1016/j.cej.2024.155508.

[130] Krishnan, A., Dujardin, E., Ebbesen, T. W., Yianilos, P. N., and Treacy, M. M. J. (1998). Young's modulus of single-walled nanotubes, *Physical Review B* **58**, 20, pp. 14013–14019, doi:10.1103/physrevb. 58.14013.

[131] Krishnan, A., Dujardin, E., Treacy, M. M. J., Hugdahl, J., Lynum, S., and Ebbesen, T. W. (1997). Graphitic cones and the nucleation of curved carbon surfaces, *Nature* **388**, 6641, pp. 451–454, doi:10. 1038/41284.

[132] Krive, I. V., Shekhter, R. I., and Jonson, M. (2006). Carbon "peapods"—a new tunable nanoscale graphitic structure (review), *Low Temperature Physics* **32**, 10, pp. 887–905, doi:10.1063/1. 2364474.

[133] Krivtsov, A. M. and Podol'skaya, E. A. (2010). Modeling of elastic properties of crystals with hexagonal close-packed lattice, *Mechanics of Solids* **45**, 3, pp. 370–378, doi:10.3103/s0025654410030076.

[134] Kroto, H. (1997). Symmetry, space, stars, and C_{60} (Nobel lecture), *Angewandte Chemie International Edition in English* **36**, 15, pp. 1578–1593, doi:10.1002/anie.199715781.

[135] Kroto, H. W., Heath, J. R., O'Brien, S. C., Curl, R. F., and Smalley, R. E. (1985). C_{60}: Buckminsterfullerene, *Nature* **318**, 6042, pp. 162–163, doi:10.1038/318162a0.

[136] Krylova, K. A., Safina, L. R., Rusalev, Y. V., and Baimova, J. A. (2025). The impact of composite graphene/Ni coating on nanoindentation of Ni: Deformation mechanism, *Surfaces and Interfaces* **61**, 15, p. 106093.

[137] Krylova, K. A., Baimova, J. A., Lobzenko, I. P., and Rudskoy, A. I. (2020). Crumpled graphene as a hydrogen storage media: Atomistic simulation, *Physica B: Condensed Matter* **583**, p. 412020, doi:10. 1016/j.physb.2020.412020.

[138] Krylova, K. A., Safina, L. R., Shcherbinin, S. A., and Baimova, J. A. (2022). Methodology and for molecular dynamics simulation of plastic deformation of a nickel/graphene composite, *Materials* **15**, 11, p. 4038, doi:10.3390/ma15114038.

[139] Kuznetsov, V., Chuvilin, A., Moroz, E., Kolomiichuk, V., Shaikhutdinov, S., Butenko, Y., and Mal'kov, I. (1994). Effect of explosion conditions on the structure of detonation soots: Ultradisperse diamond and onion carbon, *Carbon* **32**, 5, pp. 873–882, doi:10.1016/0008-6223(94)90044-2.

[140] Kuznetsov, V. L., Chuvilin, A. L., Butenko, Y. V., Mal'kov, I. Y., and Titov, V. M. (1994). Onion-like carbon from ultra-disperse diamond, *Chemical Physics Letters* **222**, 4, pp. 343–348, doi:10.1016/0009-2614(94)87072-1.

[141] Kvashnin, A. G. and Sorokin, P. B. (2014). Lonsdaleite films with nanometer thickness, *The Journal of Physical Chemistry Letters* **5**, 3, pp. 541–548, doi:10.1021/jz402528q.

[142] L.D., L. (1937). Towards the theory of phase transitions. ii. *Phys. Z. Sowjetunion* **11**, p. 545.

[143] Land, T., Michely, T., Behm, R., Hemminger, J., and Comsa, G. (1992). STM investigation of single layer graphite structures produced on Pt(111) by hydrocarbon decomposition, *Surface Science* **264**, 3, pp. 261–270, doi:10.1016/0039-6028(92)90183-7.

[144] Lavini, F., Rejhon, M., and Riedo, E. (2022). Two-dimensional diamonds from sp2-to-sp3 phase transitions, *Nature Reviews Materials* **7**, 10, pp. 814–832, doi:10.1038/s41578-022-00451-y.

[145] Lee, C., Wei, X., Kysar, J. W., and Hone, J. (2008). Measurement of the elastic properties and intrinsic strength of monolayer graphene, *Science* **321**, 5887, pp. 385–388, doi:10.1126/science.1157996.

[146] Leenaerts, O., Partoens, B., and Peeters, F. M. (2009). Hydrogenation of bilayer graphene and the formation of bilayer graphane from first principles, *Physical Review B* **80**, 24, p. 245422, doi:10.1103/physrevb.80.245422.

[147] Lei, J. and Liu, Z. (2018). The structural and mechanical properties of graphene aerogels based on Schwarz-surface-like graphene models, *Carbon* **130**, pp. 741–748, doi:10.1016/j.carbon.2018.01.061.

[148] Li, G., Li, Y., Liu, H., Guo, Y., Li, Y., and Zhu, D. (2010). Architecture of graphdiyne nanoscale films, *Chemical Communications* **46**, 19, p. 3256, doi:10.1039/b922733d.

[149] Li, G., Li, Y., Qian, X., Liu, H., Lin, H., Chen, N., and Li, Y. (2011). Construction of tubular molecule aggregations of graphdiyne for highly efficient field emission, *The Journal of Physical Chemistry C* **115**, 6, pp. 2611–2615, doi:10.1021/jp107996f.

[150] Li, M., Deng, T., Zheng, B., Zhang, Y., Liao, Y., and Zhou, H. (2019). Effect of defects on the mechanical and thermal properties of graphene, *Nanomaterials* **9**, 3, p. 347, doi:10.3390/nano9030347.

[151] Lisovenko, D. S., Baimova, J. A., Rysaeva, L. K., Gorodtsov, V. A., Rudskoy, A. I., and Dmitriev, S. V. (2016). Equilibrium diamond-like carbon nanostructures with cubic anisotropy: Elastic properties, *physica status solidi (b)* **253**, 7, pp. 1295–1302, doi:10.1002/pssb.201600049.

[152] Liu, B., Gao, T., Liao, P., Wen, Y., Yao, M., Shi, S., and Zhang, W. (2021). Metallic VS_2 graphene heterostructure as an ultra-high rate and high-specific capacity anode material for Li/Na-ion batteries, *Physical Chemistry Chemical Physics* **23**, 34, pp. 18784–18793, doi:10.1039/d1cp02243a.

[153] Liu, B., Reddy, C. D., Jiang, J., Baimova, J. A., Dmitriev, S. V., Nazarov, A. A., and Zhou, K. (2012). Morphology and in-plane thermal conductivity of hybrid graphene sheets, *Applied Physics Letters* **101**, 21, p. 211909, doi:10.1063/1.4767388.

[154] Liu, C., Xu, X., Qiu, L., Wu, M., Qiao, R., Wang, L., Wang, J., Niu, J., Liang, J., Zhou, X., Zhang, Z., Peng, M., Gao, P., Wang, W., Bai, X., Ma, D., Jiang, Y., Wu, X., Yu, D., Wang, E., Xiong, J., Ding, F., and Liu, K. (2019). Kinetic modulation of graphene growth by fluorine through spatially confined decomposition of metal fluorides, *Nature Chemistry* **11**, 8, pp. 730–736, doi:10.1038/s41557-019-0290-1.

[155] Liu, K., Yan, Q., Chen, M., Fan, W., Sun, Y., Suh, J., Fu, D., Lee, S., Zhou, J., Tongay, S., Ji, J., Neaton, J. B., and Wu, J. (2014). Elastic properties of chemical-vapor-deposited monolayer MoS_2, WS_2, and their bilayer heterostructures, *Nano Letters* **14**, 9, pp. 5097–5103, doi:10.1021/nl501793a.

[156] Liu, P., Li, X., Min, P., Chang, X., Shu, C., Ding, Y., and Yu, Z.-Z. (2020). 3D lamellar-structured graphene aerogels for thermal interface composites with high through-plane thermal conductivity and fracture toughness, *Nano-Micro Letters* **13**, 1, p. 22, doi:10.1007/s40820-020-00548-5.

[157] Liu, T.-H., Gajewski, G., Pao, C.-W., and Chang, C.-C. (2011). Structure, energy, and structural transformations of graphene grain boundaries from atomistic simulations, *Carbon* **49**, 7, pp. 2306–2317, doi:10.1016/j.carbon.2011.01.063.

[158] Liu, Y. and Yakobson, B. I. (2010). Cones, pringles, and grain boundary landscapes in graphene topology, *Nano Letters* **10**, 6, pp. 2178–2183, doi:10.1021/nl100988r.

[159] Lobzenko, I., Baimova, J., and Krylova, K. (2020). Hydrogen on graphene with low amplitude ripples: First-principles calculations, *Chemical Physics* **530**, p. 110608, doi:10.1016/j.chemphys.2019.110608.

[160] Lu, Q. and Huang, R. (2009). Nonlinear mechanics of single-atomic-layer graphene sheets, *International Journal of Applied Mechanics* **01**, 03, pp. 443–467, doi:10.1142/s1758825109000228.

[161] Luu, H.-T., Dang, S.-L., Hoang, T.-V., and Gunkelmann, N. (2021). Molecular dynamics simulation of nanoindentation in Al and Fe: on the influence of system characteristics, *Applied Surface Science* **551**, p. 149221, doi:10.1016/j.apsusc.2021.149221.

[162] Lv, Z., Mao, Y., Zhang, Q., Liu, Y., and Li, R. (2023). Molecular dynamics analysis on the indentation hardness of nano-twinned nickel, *Materials Today Communications* **37**, p. 107313, doi:10.1016/j.mtcomm.2023.107313.

[163] Malko, D., Neiss, C., Viñes, F., and Görling, A. (2012). Competition for graphene: Graphynes with direction-dependent Dirac cones, *Physical Review Letters* **108**, 8, p. 086804, doi:10.1103/physrevlett.108.086804.

[164] Manzetti, S. (2013). Molecular and crystal assembly inside the carbon nanotube: encapsulation and manufacturing approaches, *Advances in Manufacturing* **1**, 3, pp. 198–210, doi:10.1007/s40436-013-0030-5.

[165] Mendelev, M. and King, A. (2013). The interactions of self-interstitials with twin boundaries, *Philosophical Magazine* **93**, 10–12, pp. 1268–1278, doi:10.1080/14786435.2012.747012.

[166] Mendelev, M., Kramer, M., Becker, C., and Asta, M. (2008). Analysis of semi-empirical interatomic potentials appropriate for simulation of crystalline and liquid Al and Cu, *Philosophical Magazine* **88**, 12, pp. 1723–1750, doi:10.1080/14786430802206482.

[167] Mendelev, M., Kramer, M., Hao, S., Ho, K., and Wang, C. (2012). Development of interatomic potentials appropriate for simulation of liquid and glass properties of NiZr$_2$ alloy, *Philosophical Magazine* **92**, 35, pp. 4454–4469, doi:10.1080/14786435.2012.712220.

[168] Mendelev, M. I., Underwood, T. L., and Ackland, G. J. (2016). Development of an interatomic potential for the simulation of defects, plasticity, and phase transformations in titanium, *The Journal of Chemical Physics* **145**, 15, p. 154102, doi:10.1063/1.4964654.

[169] Mermin, N. D. (1968). Crystalline order in two dimensions, *Physical Review* **176**, 1, pp. 250–254, doi:10.1103/physrev.176.250.

[170] Meyer, J. C., Geim, A. K., Katsnelson, M. I., Novoselov, K. S., Booth, T. J., and Roth, S. (2007). The structure of suspended graphene sheets, *Nature* **446**, 7131, pp. 60–63, doi:10.1038/nature05545.

[171] Miao, J. and Fan, T. (2023). Flexible and stretchable transparent conductive graphene-based electrodes for emerging wearable electronics, *Carbon* **202**, pp. 495–527, doi:10.1016/j.carbon.2022.11.018.

[172] Miranda, R. and Vázquez de Parga, A. L. (2009). Surfing ripples towards new devices, *Nature Nanotechnology* **4**, 9, pp. 549–550, doi: 10.1038/nnano.2009.250.

[173] Moessinger, D., Chaudhuri, D., Kudernac, T., Lei, S., De Feyter, S., Lupton, J. M., and Hooger, S. (2010). Large all-hydrocarbon spoked wheels of high symmetry: Modular synthesis, photophysical properties, and surface assembly, *Journal of the American Chemical Society* **132**, 4, pp. 1410–1423, doi:10.1021/ja909229y.

[174] Mohiuddin, T. M. G., Lombardo, A., Nair, R. R., Bonetti, A., Savini, G., Jalil, R., Bonini, N., Basko, D. M., Galiotis, C., Marzari, N., Novoselov, K. S., Geim, A. K., and Ferrari, A. C. (2009). Uniaxial strain in graphene by Raman spectroscopy: G peak splitting, Grüneisen parameters, and sample orientation, *Physical Review B* **79**, 20, p. 205433, doi:10.1103/physrevb.79.205433.

[175] Mohr, M., Maultzsch, J., Dobardžić, E., Reich, S., Milošević, I., Damnjanović, M., Bosak, A., Krisch, M., and Thomsen, C. (2007). Phonon dispersion of graphite by inelastic x-ray scattering. *Physical Review B* **76**, p. 035439, doi:10.1103/PhysRevB.76.035439.

[176] Montblanch, A. R.-P., Barbone, M., Aharonovich, I., Atatüre, M., and Ferrari, A. C. (2023). Layered materials as a platform for quantum technologies, *Nature Nanotechnology* **18**, 6, pp. 555–571, doi:10.1038/s41565-023-01354-x.

[177] Mortazavi, B., Shojaei, F., Javvaji, B., Azizi, M., Zhan, H., Rabczuk, T., and Zhuang, X. (2020). First-principles investigation of mechanical, electronic and optical properties of H-, F-, and Cl-diamane, *Applied Surface Science* **528**, p. 147035, doi:10.1016/j.apsusc.2020. 147035.

[178] Mouhat, F. and Coudert, F.-X. (2014). Necessary and sufficient elastic stability conditions in various crystal systems, *Physical Review B* **90**, 22, p. 224104, doi:10.1103/physrevb.90.224104.

[179] Mubeen, S., Zhang, T., Yoo, B., Deshusses, M. A., and Myung, N. V. (2007). Palladium nanoparticles decorated single-walled carbon nanotube hydrogen sensor, *Journal of Physical Chemistry C* **111**, 17, pp. 6321–6327, doi:10.1021/jp067716m.

[180] Muniz, A. R. and Maroudas, D. (2012). Formation of fullerene superlattices by interlayer bonding in twisted bilayer graphene, *Journal of Applied Physics* **111**, 4, p. 043513, doi:10.1063/1.3682475.

[181] Muniz, A. R. and Maroudas, D. (2012). Opening and tuning of band gap by the formation of diamond superlattices in twisted bilayer graphene, *Physical Review B* **86**, 7, p. 075404, doi:10.1103/physrevb. 86.075404.

[182] Muniz, A. R. and Maroudas, D. (2013). Superlattices of fluorinated interlayer-bonded domains in twisted bilayer graphene, *The Journal of Physical Chemistry C* **117**, 14, pp. 7315–7325, doi:10.1021/ jp310184c.

[183] Nagashima, A., Nuka, K., Itoh, H., Ichinokawa, T., Oshima, C., and Otani, S. (1993). Electronic states of monolayer graphite formed on TiC(111) surface, *Surface Science* **291**, 1–2, pp. 93–98, doi:10.1016/ 0039-6028(93)91480-d.

[184] Neek-Amal, M. and Peeters, F. M. (2010). Graphene nanoribbons subjected to axial stress, *Physical Review B* **82**, 8, p. 085432, doi: 10.1103/physrevb.82.085432.

[185] Nelson, D. and Peliti, L. (1987). Fluctuations in membranes with crystalline and hexatic order, *Journal de Physique* **48**, 7, pp. 1085– 1092, doi:10.1051/jphys:019870048070108500.

[186] Nelson, M. T., Humphrey, W., Gursoy, A., Dalke, A., Kalé, L. V., Skeel, R. D., and Schulten, K. (1996). Namd: a parallel, object-oriented molecular dynamics program, *The International Journal of Supercomputer Applications and High Performance Computing* **10**, 4, pp. 251–268, doi:10.1177/109434209601000401.

[187] Ni, Z. H., Yu, T., Lu, Y. H., Wang, Y. Y., Feng, Y. P., and Shen, Z. X. (2008). Uniaxial strain on graphene: Raman spectroscopy study and band-gap opening, *ACS Nano* **2**, 11, pp. 2301–2305, doi:10.1021/ nn800459e.

[188] Novikov, I. V., Raginov, N. I., Krasnikov, D. V., Zhukov, S. S., Zhivetev, K. V., Terentiev, A. V., Ilatovskii, D. A., Elakshar, A., Khabushev, E. M., Grebenko, A. K., Kuznetsov, S. A., Shandakov, S. D., Gorshunov, B. P., and Nasibulin, A. G. (2024). Fast liquid-free patterning of SWCNT films for electronic and optical applications, *Chemical Engineering Journal* **485**, p. 149733, doi:10.1016/j.cej. 2024.149733.

[189] Novoselov, K. S., Geim, A. K., Morozov, S. V., Jiang, D., Zhang, Y., Dubonos, S. V., Grigorieva, I. V., and Firsov, A. A. (2004). Electric field effect in atomically thin carbon films, *Science* **306**, 5696, pp. 666–669, doi:10.1126/science.1102896.

[190] Okada, S., Otani, M., and Oshiyama, A. (2003). Electron-state control of carbon nanotubes by space and encapsulated fullerenes, *Physical Review B* **67**, 20, p. 205411, doi:10.1103/physrevb.67. 205411.

[191] Olabi, A., Abdelkareem, M. A., Wilberforce, T., and Sayed, E. T. (2021). Application of graphene in energy storage device – a review, *Renewable and Sustainable Energy Reviews* **135**, p. 110026, doi:10. 1016/j.rser.2020.110026.

[192] Oliveira, T. A., Silva, P. V., Saraiva-Souza, A., da Silva Filho, J. G., and Girão, E. C. (2022). Structural and electronic properties of nonconventional alpha -graphyne nanocarbons, *Physical Review Materials* **6**, 1, p. 016001, doi:10.1103/physrevmaterials.6.016001.

[193] Oliver, W. and Pharr, G. (1992). An improved technique for determining hardness and elastic modulus using load and displacement sensing indentation experiments, *Journal of Materials Research* **7**, 6, pp. 1564–1583, doi:10.1557/jmr.1992.1564.

[194] Oshima, C. and Nagashima, A. (1997). Ultra-thin epitaxial films of graphite and hexagonal boron nitride on solid surfaces, *Journal of Physics: Condensed Matter* **9**, 1, pp. 1–20, doi:10.1088/0953-8984/ 9/1/004.

[195] Ou, B., Yan, J., Wang, Q., and Lu, L. (2022). Thermal conductance of graphene-titanium interface: A molecular simulation, *Molecules* **27**, 3, p. 905, doi:10.3390/molecules27030905.

[196] Pak, J. J., Weakley, T. J. R., and Haley, M. M. (1999). Stepwise assembly of site specifically functionalized dehydrobenzo[18] annulenes, *Journal of the American Chemical Society* **121**, 36, pp. 8182–8192, doi:10.1021/ja991749g.

[197] Pakornchote, T., Ektarawong, A., Alling, B., Pinsook, U., Tancharakorn, S., Busayaporn, W., and Bovornratanaraks, T. (2019). Phase stabilities and vibrational analysis of hydrogenated diamondized bilayer graphenes: A first principles investigation, *Carbon* **146**, pp. 468–475, doi:10.1016/j.carbon.2019.01.088.

[198] Pastorelli, R., Ferrari, A., Beghi, M., Bottani, C., and Robertson, J. (2000). Elastic constants of ultrathin diamond-like carbon films, *Diamond and Related Materials* **9**, 3–6, pp. 825–830, doi:10.1016/ s0925-9635(99)00245-9.

[199] Patil, S. P., Rege, A., Sagardas, Itskov, M., and Markert, B. (2017). Mechanics of nanostructured porous silica aerogel resulting from molecular dynamics simulations, *The Journal of Physical Chemistry B* **121**, 22, pp. 5660–5668, doi:10.1021/acs.jpcb.7b03184.

[200] Patil, S. P., Shendye, P., and Markert, B. (2020). Molecular investigation of mechanical properties and fracture behavior of graphene

aerogel, *The Journal of Physical Chemistry B* **124**, 28, pp. 6132–6139, doi:10.1021/acs.jpcb.0c03977.

[201] Peelaers, H., Hernández-Nieves, A. D., Leenaerts, O., Partoens, B., and Peeters, F. M. (2011). Vibrational properties of graphene fluoride and graphane, *Applied Physics Letters* **98**, 5, p. 051914, doi:10.1063/1.3551712.

[202] Peierls, R. (1934). Bemerkungen uber umwandlungstemperaturen, *Helv. Phys. Acta* **7**, p. 81.

[203] Pekala, R., Alviso, C., Kong, F., and Hulsey, S. (1992). Aerogels derived from multifunctional organic monomers, *Journal of Non-Crystalline Solids* **145**, pp. 90–98, doi:10.1016/S0022-3093(05)80436-3.

[204] Peng, Q., Ji, W., and De, S. (2012). Mechanical properties of graphyne monolayers: a first-principles study, *Physical Chemistry Chemical Physics* **14**, 38, p. 13385, doi:10.1039/c2cp42387a.

[205] Peng, Q., Liang, C., Ji, W., and De, S. (2013). A theoretical analysis of the effect of the hydrogenation of graphene to graphane on its mechanical properties, *Phys. Chem. Chem. Phys.* **15**, 6, pp. 2003–2011, doi:10.1039/c2cp43360e.

[206] Pham, K. D. (2022). Tunable structural and electronic properties of C_4XY (X = Y = H, Cl and F) monolayers by functionalization, electric field and strain engineering, *New Journal of Chemistry* **46**, 19, pp. 9383–9388, doi:10.1039/d2nj01076c.

[207] Piazza, F., Cruz, K., Monthioux, M., Puech, P., and Gerber, I. (2020). Raman evidence for the successful synthesis of diamane, *Carbon* **169**, pp. 129–133, doi:10.1016/j.carbon.2020.07.068.

[208] Piazza, F., Monthioux, M., Puech, P., Gerber, I. C., and Gough, K. (2021). Progress on diamane and diamanoid thin film pressureless synthesis, *C* **7**, 1, p. 9, doi:10.3390/c7010009.

[209] Plimpton, S. (1995). Fast parallel algorithms for short-range molecular dynamics, *Journal of Computational Physics* **117**, 1, pp. 1–19, doi:10.1006/jcph.1995.1039.

[210] Plummer, G. and Tucker, G. J. (2019). Bond-order potentials for the Ti_3AlC_2 and Ti_3SiC_2 MAX phases, *Physical Review B* **100**, 21, p. 214114, doi:10.1103/physrevb.100.214114.

[211] Podlivaev, A. I. and Katin, K. P. (2025). Competition of hydrogen desorption and migration on graphene surface in alternating electric field: Multiscale molecular dynamics and diffusion study, *Applied Surface Science* **686**, p. 162125, doi:10.1016/j.apsusc.2024.162125.

[212] Polyakova, P. V., Murzaev, R. T., Lisovenko, D. S., and Baimova, J. A. (2024). Elastic constants of graphane, graphyne, and graphdiyne, *Computational Materials Science* **244**, p. 113171, doi:10.1016/j.commatsci.2024.113171.

[213] Polyakova, P., Galiakhmetova, L., Murzaev, R., Lisovenko, D., and Baimova, J. (2023). Elastic properties of diamane, *Letters on Materials* **13**, 2, pp. 171–176, doi:10.22226/2410-3535-2023-2-171-176.

[214] Polyakova, P. V. and Murzaev, R. T. (2023). Methodology for calculation of elastic constants of diamane by molecular dynamics, in *2023 IEEE 24th International Conference of Young Professionals in Electron Devices and Materials (EDM)* (IEEE), pp. 60–64, doi: 10.1109/edm58354.2023.10225206.

[215] Polyakova, P. V., Murzaev, R. T., and Baimova, J. A. (2025). Mechanical properties of diamane: Orientation dependence of strength and fracture strain, *Applied Surface Science* **681**, p. 161441, doi:10.1016/j.apsusc.2024.161441.

[216] Poot, M. and van der Zant, H. S. J. (2008). Nanomechanical properties of few-layer graphene membranes, doi:10.48550/ARXIV. 0802.0413.

[217] Proctor, J. E., Armada, D. M., and Vijayaraghavan, A. (2017). *An Introduction to Graphene and Carbon Nanotubes* (Taylor and Francis Group).

[218] Qin, H., Sun, Y., Liu, J. Z., Li, M., and Liu, Y. (2017). Negative poisson's ratio in rippled graphene, *Nanoscale* **9**, 12, pp. 4135–4142, doi:10.1039/c6nr07911c.

[219] Qiu, D., Wang, Q., Cheng, S., Gao, N., and Li, H. (2019). Electronic structures of two-dimensional hydrogenated bilayer diamond films with Si dopant and Si-V center, *Results in Physics* **13**, p. 102240, doi:10.1016/j.rinp.2019.102240.

[220] Qiu, L., Huang, B., He, Z., Wang, Y., Tian, Z., Liu, J. Z., Wang, K., Song, J., Gengenbach, T. R., and Li, D. (2017). Extremely low density and super-compressible graphene cellular materials, *Advanced Materials* **29**, 36, p. 1701553, doi:10.1002/adma.201701553.

[221] Qiu, L., Liu, J. Z., Chang, S. L., Wu, Y., and Li, D. (2012). Biomimetic superelastic graphene-based cellular monoliths, *Nature Communications* **3**, 1, p. 1241, doi:10.1038/ncomms2251.

[222] Monthioux, M. and Kuznetsov, V. L. (2006). Who should be given the credit for the discovery of carbon nanotubes?, *Carbon* **44**, 9, pp. 1621–1623, doi:10.1016/j.carbon.2006.03.019.

[223] Raeisi, M., Mortazavi, B., Podryabinkin, E. V., Shojaei, F., Zhuang, X., and Shapeev, A. V. (2020). High thermal conductivity in semiconducting Janus and non-Janus diamanes, *Carbon* **167**, pp. 51–61, doi:10.1016/j.carbon.2020.06.007.

[224] Rahman, M. H., Mitra, S., Motalab, M., and Bose, P. (2020). Investigation on the mechanical properties and fracture phenomenon

of silicon doped graphene by molecular dynamics simulation, *RSC Advances* **10**, 52, pp. 31318–31332, doi:10.1039/d0ra06085b.

[225] Rajasekaran, S., Abild-Pedersen, F., Ogasawara, H., Nilsson, A., and Kaya, S. (2013). Interlayer carbon bond formation induced by hydrogen adsorption in few-layer supported graphene, *Physical Review Letters* **111**, 8, p. 085503, doi:10.1103/physrevlett.111.085503.

[226] Riaz, M. A., Hadi, P., Abidi, I. H., Tyagi, A., Ou, X., and Luo, Z. (2017). Recyclable 3D graphene aerogel with bimodal pore structure for ultrafast and selective oil sorption from water, *RSC Advances* **7**, 47, pp. 29722–29731, doi:10.1039/c7ra02886e.

[227] Robertson, J. (2002). Diamond-like amorphous carbon, *Materials Science and Engineering: R: Reports* **37**, 4–6, pp. 129–281, doi: 10.1016/s0927-796x(02)00005-0.

[228] Rochefort, A. (2003). Electronic and transport properties of carbon nanotube peapods, *Physical Review B* **67**, 11, p. 115401, doi:10.1103/physrevb.67.115401.

[229] Rojas, M. I. and Leiva, E. P. M. (2007). Density functional theory study of a graphene sheet modified with titanium in contact with different adsorbates, *Physical Review B* **76**, 15, p. 155415, doi:10.1103/physrevb.76.155415.

[230] Rozhnova, E. and Baimova, J. (2024). Morphology of graphene aerogel as the key factor: Mechanical properties under tension and compression, *Gels* **11**, 1, p. 3, doi:10.3390/gels11010003.

[231] Ruestes, C., Stukowski, A., Tang, Y., Tramontina, D., Erhart, P., Remington, B., Urbassek, H., Meyers, M., and Bringa, E. (2014). Atomistic simulation of tantalum nanoindentation: effects of indenter diameter, penetration velocity, and interatomic potentials on defect mechanisms and evolution, *Materials Science and Engineering: A* **613**, pp. 390–403, doi:10.1016/j.msea.2014.07.001.

[232] Rysaeva, L. K., Baimova, J. A., Dmitriev, S. V., Lisovenko, D. S., Gorodtsov, V. A., and Rudskoy, A. I. (2019). Elastic properties of diamond-like phases based on carbon nanotubes, *Diamond and Related Materials* **97**, p. 107411, doi:10.1016/j.diamond.2019.04.034.

[233] Rysaeva, L. K., Lisovenko, D. S., Gorodtsov, V. A., and Baimova, J. A. (2020). Stability, elastic properties and deformation behavior of graphene-based diamond-like phases, *Computational Materials Science* **172**, p. 109355, doi:10.1016/j.commatsci.2019.109355.

[234] Safina, L. L. and Baimova, J. A. (2020). Molecular dynamics simulation of fabrication of Ni-graphene composite: temperature effect, *Micro & Nano Letters* **15**, 3, pp. 176–180, doi:10.1049/mnl.2019.0414.

[235] Safina, L. R., Baimova, J. A., and Mulyukov, R. R. (2019). Nickel nanoparticles inside carbon nanostructures: atomistic simulation, *Mechanics of Advanced Materials and Modern Processes* **5**, 1, p. 2, doi:10.1186/s40759-019-0042-3.

[236] Safina, L. R., Baimova, J. A., Krylova, K. A., Murzaev, R. T., Shcherbinin, S. A., and Mulyukov, R. R. (2021). Ni-graphene composite obtained by pressure-temperature treatment: Atomistic simulations, *physica status solidi (RRL)* **15**, 11, p. 2100429, doi: 10.1002/pssr.202100429.

[237] Safina, L. R., Krylova, K. A., Murzaev, R. T., Baimova, J. A., and Mulyukov, R. R. (2021). Crumpled graphene-storage media for hydrogen and metal nanoclusters, *Materials* **14**, 9, p. 2098, doi: 10.3390/ma14092098.

[238] Sagar, T. C., Chinthapenta, V., and Horstemeyer, M. F. (2020). Effect of defect guided out-of-plane deformations on the mechanical properties of graphene, *Fullerenes, Nanotubes and Carbon Nanostructures* **29**, 2, pp. 83–99, doi:10.1080/1536383x.2020.1813720.

[239] Sakamoto, J., Heijst, J. v., Lukin, O., and Dieter Schlüter, A. (2009). Two-dimensional polymers: Just a dream of synthetic chemists? *Angewandte Chemie International Edition* **48**, 6, pp. 1030–1069, doi: 10.1002/anie.200801863.

[240] Sakhaee-Pour, A. (2009). Elastic properties of single-layered graphene sheet, *Solid State Communications* **149**, 1–2, pp. 91–95, doi:10.1016/j.ssc.2008.09.050.

[241] Samantara, A. K., Mishra, D. K., Suryawanshi, S. R., More, M. A., Thapa, R., Late, D. J., Jena, B. K., and Rout, C. S. (2015). Facile synthesis of Ag nanowire–rGO composites and their promising field emission performance, *RSC Advances* **5**, 52, pp. 41887–41893, doi: 10.1039/c5ra00308c.

[242] Samarakoon, D. K. and Wang, X.-Q. (2010). Tunable band gap in hydrogenated bilayer graphene, *ACS Nano* **4**, 7, pp. 4126–4130, doi: 10.1021/nn1007868.

[243] Sano, N., Wang, H., Alexandrou, I., Chhowalla, M., Teo, K. B. K., Amaratunga, G. A. J., and Iimura, K. (2002). Properties of carbon onions produced by an arc discharge in water, *Journal of Applied Physics* **92**, 5, pp. 2783–2788, doi:10.1063/1.1498884.

[244] Sarma, J. V. N., Chowdhury, R., and Rengaswamy, J. (2014). Graphyne-based single electron transistor: Ab initio analysis, *Nano* **09**, 03, p. 1450032, doi:10.1142/s1793292014500325.

[245] Savin, A. V. and Savina, O. I. (2004). Nonlinear dynamics of carbon molecular lattices: Soliton plane waves in graphite layers and

supersonic acoustic solitons in nanotubes, *Physics of the Solid State* **46**, pp. 383–391, doi:10.1134/1.1649441.

[246] Schur, D. V., Zaginaichenko, S., and Veziroglu, T. N. (2008). Peculiarities of hydrogenation of pentatomic carbon molecules in the frame of fullerene molecule C_{60}, *International Journal of Hydrogen Energy* **33**, 13, pp. 3330–3345.

[247] Shenderova, O. A., Zhirnov, V. V., and Brenner, D. W. (2002). Carbon nanostructures, *Critical Reviews in Solid State and Materials Sciences* **27**, 3–4, pp. 227–356, doi:10.1080/10408430208500497.

[248] Shenoy, V. B., Reddy, C. D., Ramasubramaniam, A., and Zhang, Y. W. (2008). Edge-stress-induced warping of graphene sheets and nanoribbons, *Physical Review Letters* **101**, 24, p. 245501, doi:10. 1103/physrevlett.101.245501.

[249] Shu, H. (2021). Novel Janus diamane C_4FCl: a stable and moderate bandgap semiconductor with a huge excitonic effect, *Physical Chemistry Chemical Physics* **23**, 34, pp. 18951–18957, doi:10.1039/ d1cp02632a.

[250] Shu, H. (2021). Strain effects on stability, electronic and optical properties of two-dimensional C_4X_2 (X = F, Cl, Br), *Journal of Materials Chemistry C* **9**, 13, pp. 4505–4513, doi:10.1039/d1tc00507c.

[251] Silveira, J. F. and Muniz, A. R. (2018). Diamond nanothread-based 2D and 3D materials: Diamond nanomeshes and nanofoams, *Carbon* **139**, pp. 789–800, doi:10.1016/j.carbon.2018.07.021.

[252] Simon, F., Peterlik, H., Pfeiffer, R., Bernardi, J., and Kuzmany, H. (2007). Fullerene release from the inside of carbon nanotubes: A possible route toward drug delivery, *Chemical Physics Letters* **445**, 4–6, pp. 288–292, doi:10.1016/j.cplett.2007.08.014.

[253] Singh, D., Khossossi, N., Luo, W., Ainane, A., and Ahuja, R. (2022). 2D Janus and non-Janus diamanes with an in-plane negative Poisson's ratio for energy applications, *Materials Today Advances* **14**, p. 100225, doi:10.1016/j.mtadv.2022.100225.

[254] Golberg, D., Bando, Y., Bourgeois, L., Kurashima, K., and Sato, T. (2000). Large-scale synthesis and HRTEM analysis of single-walled B- and N-doped carbon nanotube bundles, *Carbon* **38**, 14, pp. 2017–2027, doi:10.1016/S0008-6223(00)00058-0.

[255] Sladkov, A. M., Kasatochkin, V. I., Korshak, V. V., and Kudryavtsev, Y. P. (1971). *Inventor's Certification*, No. 107, Moscow, December.

[256] Sloan, J., Dunin-Borkowski, R. E., Hutchison, J. L., Coleman, K. S., Clifford Williams, V., Claridge, J. B., York, A. P., Xu, C., Bailey, S. R., Brown, G., Friedrichs, S., and Green, M. L. (2000). The size distribution, imaging and obstructing properties of C_{60}

and higher fullerenes formed within arc-grown single walled carbon nanotubes, *Chemical Physics Letters* **316**, 3–4, pp. 191–198, doi: 10.1016/s0009-2614(99)01250-6.

[257]   Smith, B. W., Monthioux, M., and Luzzi, D. E. (1998). Encapsulated C_{60} in carbon nanotubes, *Nature* **396**, 6709, pp. 323–324, doi:10. 1038/24521.

[258]   Smith, B. W., Monthioux, M., and Luzzi, D. E. (1999). Carbon nanotube encapsulated fullerenes: a unique class of hybrid materials, *Chemical Physics Letters* **315**, 1–2, pp. 31–36, doi:10.1016/ s0009-2614(99)00896-9.

[259]   Snow, T. P. and Brownlee, D. (2024). *The Sixth Element: How Carbon Shapes Our World* (Princeton University Press).

[260]   Sofo, J. O., Chaudhari, A. S., and Barber, G. D. (2007). Graphane: A two-dimensional hydrocarbon, *Physical Review B* **75**, 15, p. 153401, doi:10.1103/physrevb.75.153401.

[261]   Song, B., Schneider, G. F., Xu, Q., Pandraud, G., Dekker, C., and Zandbergen, H. (2011). Atomic-scale electron-beam sculpting of near-defect-free graphene nanostructures, *Nano Letters* **11**, 6, pp. 2247–2250, doi:10.1021/nl200369r.

[262]   Stuart, S. J., Tutein, A. B., and Harrison, J. A. (2000). A reactive potential for hydrocarbons with intermolecular interactions, *The Journal of Chemical Physics* **112**, 14, pp. 6472–6486, doi:10.1063/1. 481208.

[263]   Tahani, M. and Safarian, S. (2018). Determination of rigidities, stiffness coefficients and elastic constants of multi-layer graphene sheets by an asymptotic homogenization method, *Journal of the Brazilian Society of Mechanical Sciences and Engineering* **41**, 1, p. 3, doi:10.1007/s40430-018-1499-4.

[264]   Tersoff, J. (1988). Empirical interatomic potential for carbon, with applications to amorphous carbon, *Physical Review Letters* **61**, 25, pp. 2879–2882, doi:10.1103/physrevlett.61.2879.

[265]   Thompson, T., Falardeau, E., and Hanlon, L. (1977). The electrical conductivity and optical reflectance of graphite-SbF_5 compounds, *Carbon* **15**, 1, pp. 39–43, doi:10.1016/0008-6223(77)90072-0.

[266]   Torkaman-Asadi, M. and Kouchakzadeh, M. (2022). Atomistic simulations of mechanical properties and fracture of graphene: A review, *Computational Materials Science* **210**, p. 111457, doi:10.1016/j. commatsci.2022.111457.

[267]   Treacy, M. M. J., Ebbesen, T. W., and Gibson, J. M. (1996). Exceptionally high Young's modulus observed for individual carbon nanotubes, *Nature* **381**, 6584, pp. 678–680, doi:10.1038/381678a0.

[268] Trembecka-Wójciga, K., Sobczak, J. J., and Sobczak, N. (2023). A comprehensive review of graphene-based aerogels for biomedical applications: the impact of synthesis parameters onto material microstructure and porosity, *Archives of Civil and Mechanical Engineering* **23**, 2, p. 133, doi:10.1007/s43452-023-00650-6.

[269] Tsai, J.-L. and Tu, J.-F. (2010). Characterizing mechanical properties of graphite using molecular dynamics simulation, *Materials and Design* **31**, 1, pp. 194–199, doi:10.1016/j.matdes.2009.06.032.

[270] Ubbelohde, A. R. and Lewis F. A. (1960). *Graphite and its Crystal Compounds* (Clarendon Press Oxford).

[271] Ugarte, D. (1992). Curling and closure of graphitic networks under electron-beam irradiation, *Nature* **359**, 6397, pp. 707–709, doi:10.1038/359707a0.

[272] Dato, A. (2019). Graphene synthesized in atmospheric plasmas — A review, *Journal of Materials Research* **34**, pp. 214–230. doi:10.1557/jmr.2018.470.

[273] van Duin, A. C. T., Dasgupta, S., Lorant, F., and Goddard, W. A. (2001). Reaxff: a reactive force field for hydrocarbons, *The Journal of Physical Chemistry A* **105**, 41, pp. 9396–9409, doi:10.1021/jp004368u.

[274] Varlamova, L. A., Erohin, S. V., Larionov, K. V., and Sorokin, P. B. (2022). Diamane oxide: two-dimensional film with mixed coverage and a variety of electronic properties, *The Journal of Physical Chemistry Letters* **13**, 49, pp. 11383–11390, doi:10.1021/acs.jpclett.2c02943.

[275] Vu, T. V., Phuc, H. V., Ahmad, S., Nha, V. Q., Lanh, C. V., Rai, D. P., Kartamyshev, A. I., Pham, K. D., Nhan, L. C., and Hieu, N. N. (2021). Outstanding elastic, electronic, transport and optical properties of a novel layered material C_4F_2: first-principles study, *RSC Advances* **11**, 38, pp. 23280–23287, doi:10.1039/d1ra04065k.

[276] Wang, C., Xia, K., Wang, H., Liang, X., Yin, Z., and Zhang, Y. (2018). Advanced carbon for flexible and wearable electronics, *Advanced Materials* **31**, 9, p. 1801072, doi:10.1002/adma.201801072.

[277] Wang, R.-N., Zheng, X.-H., Hao, H., and Zeng, Z. (2014). First-principles analysis of corrugations, elastic constants, and electronic properties in strained graphyne nanoribbons, *The Journal of Physical Chemistry C* **118**, 40, pp. 23328–23334, doi:10.1021/jp504534h.

[278] Wang, Y. and Shi, J. (2013). Effects of water molecules on tribological behavior and property measurements in nano-indentation processes — a numerical analysis, *Nanoscale Research Letters* **8**, 1, p. 389, doi:10.1186/1556-276x-8-389.

[279] Wang, Z. F., Zhang, Y., and Liu, F. (2011). Formation of hydrogenated graphene nanoripples by strain engineering and directed surface self-assembly, *Physical Review B* **83**, 4, p. 041403, doi: 10.1103/physrevb.83.041403.

[280] Warner, J. H., Margine, E. R., Mukai, M., Robertson, A. W., Giustino, F., and Kirkland, A. I. (2012). Dislocation-driven deformations in graphene, *Science* **337**, 6091, pp. 209–212, doi:10.1126/science.1217529.

[281] Wei, X. and Kysar, J. W. (2012). Experimental validation of multiscale modeling of indentation of suspended circular graphene membranes, *International Journal of Solids and Structures* **49**, 22, pp. 3201–3209, doi:10.1016/j.ijsolstr.2012.06.019.

[282] Wu, J.-B., Lin, M.-L., Cong, X., Liu, H.-N., and Tan, P.-H. (2018). Raman spectroscopy of graphene-based materials and its applications in related devices, *Chemical Society Reviews* **47**, 5, pp. 1822–1873, doi:10.1039/c6cs00915h.

[283] Wu, P., Du, P., Zhang, H., and Cai, C. (2015). Graphyne-supported single Fe atom catalysts for CO oxidation, *Physical Chemistry Chemical Physics* **17**, 2, pp. 1441–1449, doi:10.1039/c4cp04181j.

[284] Wu, Y., An, C., Guo, Y., Zong, Y., Jiang, N., Zheng, Q., and Yu, Z. Z. (2024). Highly aligned graphene aerogels for multifunctional composites, *Nano-Micro Letters* **16**, pp. 118, doi:10.1007/s40820-024-01357-w.

[285] Wu, Y.-C., Shao, J.-L., Zheng, Z., and Zhan, H. (2020). Mechanical properties of a single-layer diamane under tension and bending, *The Journal of Physical Chemistry C* **125**, 1, pp. 915–922, doi:10.1021/acs.jpcc.0c08172.

[286] Xie, P., Yuan, W., Liu, X., Peng, Y., Yin, Y., Li, Y., and Wu, Z. (2021). Advanced carbon nanomaterials for state-of-the-art flexible supercapacitors, *Energy Storage Materials* **36**, pp. 56–76, doi:10.1016/j.ensm.2020.12.011.

[287] Xie, Y., Cheng, T., Liu, C., Chen, K., Cheng, Y., Chen, Z., Qiu, L., Cui, G., Yu, Y., Cui, L., Zhang, M., Zhang, J., Ding, F., Liu, K., and Liu, Z. (2019). Ultrafast catalyst-free graphene growth on glass assisted by local fluorine supply, *ACS Nano* **13**, 9, pp. 10272–10278, doi:10.1021/acsnano.9b03596.

[288] Xie, Y., Xu, S., Xu, Z., Wu, H., Deng, C., and Wang, X. (2016). Interface-mediated extremely low thermal conductivity of graphene aerogel, *Carbon* **98**, pp. 381–390.

[289] Xu, B., Guo, J., Wang, X., Liu, X., and Ichinose, H. (2006). Synthesis of carbon nanocapsules containing Fe, Ni or Co by arc discharge in aqueous solution, *Carbon* **44**, 13, pp. 2631–2634.

[290]	Xu, S., Zhang, L., Wang, B., and Ruoff, R. S. (2021). Chemical vapor deposition of graphene on thin-metal films, *Cell Reports Physical Science* **2**, 3, p. 100372, doi:10.1016/j.xcrp.2021.100372.

[291]	Xu, Z. and Buehler, M. J. (2010). Interface structure and mechanics between graphene and metal substrates: a first-principles study, *Journal of Physics: Condensed Matter* **22**, 48, p. 485301, doi:10.1088/0953-8984/22/48/485301.

[292]	Xue, M., Qiu, H., and Guo, W. (2013). Exceptionally fast water desalination at complete salt rejection by pristine graphyne monolayers, *Nanotechnology* **24**, 50, p. 505720, doi:10.1088/0957-4484/24/50/505720.

[293]	Yakobson, B. I. and Avouris, P. (2001). Mechanical Properties of Carbon Nanotubes. In: Dresselhaus, M. S., Dresselhaus, G., and Avouris, P. (eds.) *Carbon Nanotubes* (Springer Berlin Heidelberg), pp. 287–327, doi:10.1007/3-540-39947-X_12.

[294]	Yan, Y., Zhou, S., and Liu, S. (2017). Atomistic simulation on nanomechanical response of indented graphene/nickel system, *Computational Materials Science* **130**, pp. 16–20, doi:10.1016/j.commatsci.2016.12.028.

[295]	Yang, D.-C., Jia, R., Wang, Y., Kong, C.-P., Wang, J., Ma, Y., Eglitis, R. I., and Zhang, H.-X. (2017). Novel carbon nanotubes rolled from 6,6,12-graphyne: Double Dirac points in 1D material, *The Journal of Physical Chemistry C* **121**, 27, pp. 14835–14844, doi:10.1021/acs.jpcc.7b01687.

[296]	Yang, Y. and Xu, X. (2012). Mechanical properties of graphyne and its family – a molecular dynamics investigation, *Computational Materials Science* **61**, pp. 83–88, doi:10.1016/j.commatsci.2012.03.052.

[297]	Yao, J., Xia, Y., Dong, S., Yu, P., and Zhao, J. (2019). Finite element analysis and molecular dynamics simulations of nanoscale crack-hole interactions in chiral graphene nanoribbons, *Engineering Fracture Mechanics* **218**, p. 106571, doi:10.1016/j.engfracmech.2019.106571.

[298]	Yildirim, T. and Ciraci, S. (2005). Titanium-decorated carbon nanotubes as a potential high-capacity hydrogen storage medium, *Physical Review Letters* **94**, 17, p. 175501, doi:10.1103/physrevlett.94.175501.

[299]	Zakharchenko, K. V., Katsnelson, M. I., and Fasolino, A. (2009). Finite temperature lattice properties of graphene beyond the quasiharmonic approximation, *Physical Review Letters* **102**, 4, p. 046808, doi:10.1103/physrevlett.102.046808.

[300] Zhang, D., Fonseca, A. F., Liang, T., Phillpot, S. R., and Sinnott, S. B. (2019). Dynamics of graphene/Al interfaces using comb3 potentials, *Physical Review Materials* **3**, 11, p. 114002, doi:10.1103/physrevmaterials.3.114002.

[301] Zhang, G., Huang, S., Wang, F., and Yan, H. (2021). Layer-dependent electronic and optical properties of 2D black phosphorus: Fundamentals and engineering, *Laser and Photonics Reviews* **15**, 6, p. 2000399, doi:10.1002/lpor.202000399.

[302] Zhang, J., Terrones, M., Park, C. R., Mukherjee, R., Monthioux, M., Koratkar, N., Kim, Y. S., Hurt, R., Frackowiak, E., Enoki, T., Chen, Y., Chen, Y., and Bianco, A. (2016). Carbon science in 2016: Status, challenges and perspectives, *Carbon* **98**, pp. 708–732, doi: 10.1016/j.carbon.2015.11.060.

[303] Zhang, T., Li, X., and Gao, H. (2014). Defects controlled wrinkling and topological design in graphene, *Journal of the Mechanics and Physics of Solids* **67**, pp. 2–13, doi:10.1016/j.jmps.2014.02.005.

[304] Zhang, Y.-Y., Pei, Q.-X., Mai, Y.-W., and Gu, Y.-T. (2014). Temperature and strain-rate dependent fracture strength of graphynes, *Journal of Physics D: Applied Physics* **47**, 42, p. 425301, doi: 10.1088/0022-3727/47/42/425301.

[305] Zhou, C., Chen, S., Lou, J., Wang, J., Yang, Q., Liu, C., Huang, D., and Zhu, T. (2014). Graphene's cousin: the present and future of graphane, *Nanoscale Research Letters* **9**, 1, p. 26, doi:10.1186/1556-276x-9-26.

[306] Zhou, K. and Liu, B. (2022). Application of molecular dynamics simulation in mechanical problems, in *Molecular Dynamics Simulation* (Elsevier), pp. 129–181, doi:10.1016/b978-0-12-816419-8.00010-6.

[307] Zhou, L., Fu, H., Lv, T., Wang, C., Gao, H., Li, D., Deng, L., and Xiong, W. (2020). Nonlinear optical characterization of 2D materials, *Nanomaterials* **10**, 11, p. 2263, doi:10.3390/nano10112263.

[308] Zhou, X. W., Ward, D. K., and Foster, M. E. (2015). An analytical bond-order potential for carbon, *Journal of Computational Chemistry* **36**, 23, pp. 1719–1735, doi:10.1002/jcc.23949.

[309] Zhu, F., Leng, J., Guo, Z., and Chang, T. (2020). Size-dependent mechanical properties of twin graphene, *Proceedings of the Institution of Mechanical Engineers, Part N: Journal of Nanomaterials, Nanoengineering and Nanosystems* **235**, 1–2, pp. 4–11, doi:10.1177/2397791420972553.

[310] Zhu, L., Li, W., and Ding, F. (2019). Giant thermal conductivity in diamane and the influence of horizontal reflection symmetry on phonon scattering, *Nanoscale* **11**, 10, pp. 4248–4257, doi:10.1039/c8nr08493a.

[311] Zhu, S. and Li, T. (2014). Hydrogenation-assisted graphene origami and its application in programmable molecular mass uptake, storage, and release, *ACS Nano* **8**, 3, pp. 2864–2872, doi:10.1021/nn500025t.

[312] Zhu, Y., Zhang, Y. C., Qi, S. H., and Xiang, Z. (2014). Molecular dynamics study on the impact of the cutting depth to the titanium nanometric cutting, *Applied Mechanics and Materials* **536–537**, pp. 1431–1434, doi:10.4028/www.scientific.net/amm.536-537.1431.

Index